HSE CONTRACT RESEARCH REPORT No. 42/1992

SICK BUILDING SYNDROME: A REVIEW OF THE EVIDENCE ON CAUSES AND SOLUTIONS

G J Raw

Building Research Establishment
Garston
Watford WD2 7JR

With acknowledgement to material provided in private communication and reports to BRE from:
Anthony Slater (Building Research Establishment) on lighting
David Tong (Building Use Studies) on management
David Lush (Ove Arup Partnership) on building services
and to
Jim Sykes (Health and Safety Executive) for his review of the literature to 1987

This report is written for those who have responsibility for buildings, building services and office staff. It seeks to define sick building syndrome (SBS), distinguishing it from other building-related health issues; to explain the importance of SBS; to summarise what is known about possible causes and to give some initial guidance as to what can be done: what design factors need to be considered and what approach should be taken in existing cases of SBS.

This report and the work it describes were funded by the Health and Safety Executive. Its contents, including any opinions and/or conclusions expressed, are those of the authors alone and do not necessarily reflect HSE policy.

SICK BUILDING SYNDROME: A Review of the evidence on causes and solutions

G J Raw

CONTENTS

1. WHAT IS SICK BUILDING SYNDROME?

1.1 SUMMARY

Sick building syndrome is a phenomenon whereby people experience a range of symptoms when in specific buildings. The symptoms are irritation of the eyes, nose, throat and skin, together with headache, lethargy, irritability and lack of concentration. Although present generally in the population, these symptoms are more prevalent in some buildings than in others, and disappear over hours or days when the afflicted person leaves the building concerned. The cause (or causes) are at present not clearly identified, but the syndrome can be discriminated from other building-related problems such as physical discomfort, infections and long-term cumulative chemical hazards such as asbestos and radon.

1.2 DEFINITION

In recent years, sick building syndrome (SBS) has emerged as a significant problem in the workplace, not only in the UK but most European countries, the USA, Canada, Australia and Japan. In the UK it is generally reckoned to be a comparatively recent problem, most reports having been published since 1980, although there were early warnings in British research in the 1960s (Black et al 1966). In other countries, especially in North America and Scandinavia, the problem was first reported some 30 years ago, both in the workplace and the home, although the term 'SBS' is relatively recent.

There is some variation in the words used to describe the phenomenon (within and between nations) for example 'building sickness', 'sick office syndrome', 'tight building syndrome', 'office eye syndrome' and other terms have been used. None fully describes the condition but 'sick building syndrome' has been accorded recognition by the World Health Organisation (WHO 1982) and is the most widely used description.

In spite of the variety of terms used, there is a clear consistency in the syndrome being described. There has, however, been a degree of variability in the definitions offered. For example, a European Communities Report (Molina et al 1989) described SBS as "a set of varied symptoms experienced predominantly by people working in air conditioned buildings .....". This follows quite closely from World Health Organisation statements (WHO 1986). The symptoms are:

- irritated, dry or watering eyes (sometimes described as itching, tiredness, smarting, redness, burning, difficulty wearing contact lenses);
- irritated, runny or blocked nose (sometimes described as congestion, nosebleeds, itchy or stuffy nose);
- dry or sore throat (sometimes described as irritation, oropharangeal symptoms, upper airway irritation, difficulty swallowing);
- dryness, itching or irritation of the skin, occasionally with rash (or specific clinical terms such as erythema, rosacea, urticaria, pruritis, xerodermia; sometimes assessed by use of moisturiser or lipstick);
- less specific symptoms such as headache, lethargy, irritability and poor concentration.

The definition adopted in this report follows from the above description, but with two caveats. First, SBS is probably most often, but not exclusively, described in places of work such as office buildings and this report is concerned primarily with office buildings. Although evidence is taken from research on other building types, there is insufficient evidence to compile a report on SBS in other types of building at this stage. Second, the reference to air-conditioned buildings is accurate in the sense that SBS is more common in such buildings, but is not appropriate as part of a definition since other buildings can also be affected.

SBS is thus defined here in terms of a group of symptoms which people experience while they are in specific buildings. The symptoms are of course present in the population at large (at unknown incidence rates) but they are distinguished by being more prevalent, as a group, in some buildings in comparison with others. Thus, to say that SBS is a real phenomenon is to say that there is a variation in symptoms among buildings, not a clear division into 'sick' and 'healthy' buildings but a continuous variation. It is not necessary that everyone in a building should be affected before SBS is suspected: the reporting of symptoms commonly varies by a factor of ten among occupants within a building. The affected group should include people who are otherwise healthy.

An important characteristic of the symptoms is that they are reduced in intensity or disappear entirely when the person is away from the suspected building. The symptoms are therefore called 'building-related'. The time required for recovery can vary from hours to days, depending on the type and severity of the symptom.

Some investigations have produced more extensive lists of symptoms, including for example, airway infections and coughs, wheezing, nausea, dizziness (WHO 1982) high blood pressure (Whorton et al 1987) and miscarriages (Ferahrian 1984). However, these are mentioned as occurring amongst staff in certain 'sick' buildings; they probably should not be attributed to SBS. Taste and odour anomalies are not necessarily symptoms: they are likely to be environmental perceptions and are therefore best excluded from the list of defining symptoms. Breathing difficulties are sometimes reported among SBS symptoms but are best used to identify individuals with asthma, rather than SBS. The range of symptoms reported, and their prevalence, will obviously depend to some extent on the number and nature of the questions used to elicit the information.

SBS is thus at present defined, as many health problems have been in the past, in terms of symptoms and conditions of occurrence, rather than cause. The reason is that there is no single proven cause, and any attempt to introduce cause into the definition is likely to be misleading at present, particularly since there are probably multiple causes. It is nevertheless possible to talk of preventive and remedial measures, much as many diseases were to some extent prevented (e.g. by hygiene practices) and treated by reducing specific symptoms (e.g. fever) long before the cause of them was identified.

Various attempts have been made to go beyond the definition adopted here. For example, SBS has been defined as a problem of indoor air quality, of 'tight' buildings, or as a syndrome "like that which would be caused by exposure to formaldehyde but in the absence of excessive formaldehyde levels". Such definitions have validity as descriptions or hypotheses, but are premature as definitions until we have a better understanding of the causes of SBS.

A particular variation on this theme is that "SBS can be diagnosed only after eliminating all other building related illnesses" (Molina et al 1989). There is an important point being made here: the term SBS is frequently applied more broadly to any building-related illness or excess of complaints from building users. This is unhelpful and it is better to maintain a distinction between SBS and, for example:

- complaints about discomfort (e.g. from temperature, noise, chairs and VDUs);
- long-term effects of identified indoor hazards (e.g. radon, asbestos);
- specific infectious illnesses caused by known organisms (e.g. Legionnaires' disease);
- building defects which do not cause SBS symptoms (e.g. structural flaws).

Such problems may occur in the same buildings, and the causal factors may overlap, but a distinction is still necessary. For example Appleby & Bailey (1990) identify a building with complaints of discomfort, largely related to air movement and environmental tobacco smoke (ETS), which was not regarded as 'sick'. SBS should be defined by a set of symptoms, not by the overall incidence of diseases or by perceived discomfort without symptoms. It is in this sense a health issue akin to allergies and asthma: the potential to react is there in the person but no symptoms appear until a particular environmental challenge is encountered.

In spite of the need to distinguish SBS from other building-related illnesses, there is some debate about whether SBS should be diagnosed only if there is no apparently obvious cause for the problems in a particular building (e.g. dampers seized in a position which prevents outside air from entering a building). There is a logic to this in practical terms: there is no point in launching into uncertain solutions to SBS if there is a clear fault (e.g. in the building or the management of the work force) which must be put right. To include this provision in the **definition** of SBS is risky because the judgement of what would constitute obvious problems is somewhat subjective and may in any case result in some major causes of SBS being under-emphasised just because they happen to fall within current knowledge. Similarly, SBS cannot be diagnosed by the observation of defects in a building or the indoor environment in the absence of data on symptoms.

Although it is now generally thought that SBS has multiple causes, there is still a tendency to refer to (and to research) single causes or classes of causes (e.g. air conditioning, indoor air quality). It is probably not helpful to react to this by **defining** SBS as having multiple causes. While it is almost certainly true that SBS has multiple causes, such a definition risks prejudging the outcome of research and may obscure the complexity of the situation: that there are probably many combinations of causes in different buildings but perhaps in some buildings a single main cause.

Some authors have attempted to subdivide SBS into different categories. Most notably, the WHO (1982) differentiates between 'temporarily sick buildings' where symptoms decrease and disappear in time and 'permanently sick buildings' where they persist, often despite the most extensive remedial measures. There is probably not such a clear distinction, but rather a range of both the complexity of causes and the difficulty of identifying the causes. Some researchers and authors have postulated subspecies of SBS based on their particular area of interest (Berglund et al 1984, Franck 1986). Such an approach may have advantages if the different symptoms of SBS can be shown to have many different causes, but is probably premature.

2. THE IMPLICATIONS OF SBS

2.1 SUMMARY

The symptoms of SBS can be regarded as minor in the sense that, apparently, no lasting physical damage is done. In fact recovery is normally reported to be quite rapid at the end of exposure. The symptoms are, however, not trivial to the people who experience them on a regular basis at their place of work and all the evidence is that the number of people affected is not trivial. Although the basis of estimates is not perfect, most figures suggest 30-50% of new or refurbished buildings are affected but slightly older buildings are affected as much if not more.

Air-conditioned buildings are generally associated with higher prevalences: in the UK approximately 55% of staff in such buildings are affected - many only mildly but to the extent that they perceive a negative effect of the office environment on their productivity. Apart from effects on productivity when staff are at work, SBS has been shown to affect absenteeism and quite obviously makes demands on the management and trades unions which spend time trying to resolve the problem. Other likely effects are on unofficial time off, reduced overtime and increased staff turnover. In extreme cases buildings may close for a period and in the long term the (probably erroneous) association of SBS with energy savings may inhibit moves towards greater energy efficiency.

2.2 PREVALENCE

The previous Section sought to define SBS. The definition of a 'sick building' is more difficult and to some extent arbitrary, depending on what is regarded as an acceptable prevalence of symptoms. For example one definition used is "a building in which complaints of ill health are more common than might reasonably be expected" (Finnegan et al 1984). A theoretical definition could follow from the definition of SBS, but in practice the identification of cases would still depend on what is regarded as an acceptable prevalence of symptoms. An analogy would be height: the height of a person can be defined and measured but this does not in itself provide a definition of 'a tall person' or 'a tall-person building'.

It is therefore difficult to define how widespread SBS is, but it is generally recognised that it is not an isolated or occasional phenomenon. Some papers suggest that perhaps 30-50% of new and remodelled buildings (with recirculating ventilation or air conditioning systems) have a high rate of complaints amongst staff and that up to 85% of staff in such buildings suffer from some symptoms (Wilson & Hedge 1987, Stellman et al 1985, Woods 1989). A World Health Organisation report (WHO 1986) concurs with this, but the estimate could be set considerably higher or lower by taking a different criterion for what would be considered a 'high rate of complaints'. These figures are particularly damning when one considers the implication that 50-70% of buildings are functioning reasonably well: apparently we can build 'healthy' buildings (i.e. with a low incidence of SBS symptoms) but we do not do it often enough.

It should also be noted that in studies of SBS the outcome is different from that used in traditional epidemiological studies because the presence of reversible symptoms at a given point or period of time is studied, rather than incidence of new cases of a disease.

There has not been a random survey to determine the actual prevalence of SBS, but probably the best available data for British buildings comes from the "Office Environment Survey" (Wilson & Hedge 1987). This was a questionnaire survey of 4373 workers in 46 office buildings which determined whether the workers had experienced any of 10 symptoms, on two or more occasions over 12 months. The percentage of people reporting each symptom is shown in Table 1.

The mean number of symptoms per person was 3.11, with a range of mean values for buildings from 1.25 to 5.25. An attempt to define a working criterion for SBS diagnosis (Raw et al 1990) specifies a building mean of more than two symptoms per person, experienced more than twice in a year. According to Wilson & Hedge (1987), 55% of respondents and all the air-conditioned buildings fitted this criterion. In practice, few studies have reported data in a form which would permit comparison with this criterion.

Table 1. PREVALENCE OF WORK-RELATED SYMPTOMS (from Wilson & Hedge, 1987)

SYMPTOM	%
Lethargy	57
Stuffy nose	47
Dry throat	46
Headache	43
Itching eyes	28*
Dry eyes	27*
Runny nose	23
Flu-like symptoms	23
Difficulty in breathing	9
Chest tightness	9

*If these two symptoms are combined, the number of people with at least one of the eye symptoms is 46 per cent.

Table 2. Average prevalence of symptoms for men (m) and women (w) at home, at work, and especially office work (from Valbjørn 1991).

Symptoms once or more times weekly	At home		At work		In Office Work			
					all people		people recovering outside the working area	
	m p.c.	w	m p.c.	w	m p.c.	w	m p.c.	w
Dryness or irritation in eyes, nose or throat	6	8	12	21	35	49	20	32
Headache, feeling heavy headed, fatigue, dizziness, malaise	-	-	-	-	28	45	26	41

A comparable survey in Denmark (see Valbjørn 1991) also showed an elevated prevalence of symptoms in office buildings, even in comparison with other work places (see Table 2).

The nearest thing to a random survey of SBS was conducted in Germany. Kröling (1987, 1988) used interviews with a representative sample of 8000 from the population of (West) Germany to obtain epidemiological data of a kind not available for the UK. The interviews were conducted at home but referred to symptoms in the workplace. The prevalences of SBS symptoms are shown in Table 3, columns (a). Other symptoms such as colds and rheumatic complaints, and discomfort in the workplace, were also more prevalent in air-conditioned buildings. In a follow-up study, respondents completed a more detailed questionnaire to describe their experience of the office environment. The results are shown in Table 3, columns (b).

Both sets of results confirm that prevalences are mostly far higher in air-conditioned buildings. The higher prevalences in the questionnaire study are typical of the difference between interview and questionnaire data, and are broadly comparable with the UK data (Wilson & Hedge 1987).

Table 3: PREVALENCE (%) OF SBS SYMPTOMS (From Kröling 1987, 1988)

	Air-conditioned buildings		Non air-conditioned buildings	
	(a)Interview	(b)Questionnaire	(a)Interview	(b)Questionnaire
Dry mucous membrane	29	57	16	31
Headache	21	39	16	26
Fatigue	20	50	15	24
Irritability	22	17	19	21
Poor concentration	14	34	10	25

Norbäck & Edling (1991) investigated the incidence of a range of symptoms in a random sample of the population of mid-Sweden. The overall prevalence of each SBS symptom was between 6% (for facial skin symptoms) and 30% (for abnormal tiredness). The relative prevalence of symptoms in different occupational groups is shown in Table 4.

There are, however some difficulties with interpreting this survey in relation to SBS. First, the sample was small for this kind of analysis: 466 respondents aged 20 - 65 completed a postal questionnaire on symptoms experienced, work factors and personal characteristics. 83% of these respondents had worked during the reference period, which was the three months prior to the survey.

Second, the symptoms were not identified as being work-related or building-related and therefore they cannot strictly be taken to be due to SBS. Another problem is that the symptoms are categorised in a way which makes them difficult to relate to SBS symptoms. 'Eye symptoms' included swollen eyelids which would not usually be listed but might be recorded as a skin symptom. 'Airway symptoms' included nasal symptoms and sore throat (which would more often be recorded as mucous membrane symptoms) and cough (not usually recorded as a separate SBS symptom). Skin symptoms included eczema and general symptoms included 'sensation of getting a cold' and nausea: the categorisations are at variance with those normally employed in SBS studies.

Third, there was no environmental monitoring but workplaces were categorised on the basis of job description according to assumed exposure to a range of pollutants. Exposure to ETS was reported by the respondent. Important workplace characteristics such as ventilation type and the number of people sharing an office were not recorded. There was no separate analysis of office workers.

Taking into account all these problems, it is difficult to evaluate the prevalence data or the correlations found with a diverse set of factors. It does, however, appear that workers doing 'office jobs' are at least as likely as other occupational groups to experience an excess of certain symptoms.

Table 4. PREVALENCE OF SYMPTOMS IN VARIOUS OCCUPATIONAL GROUPS (From Norbäck & Edling, 1991)

	Least prevalent in:	Most prevalent in:
Eye symptoms	Sales, transport, communications	Administration, management, service
Airway symptoms	Agriculture, forestry	Transport, communications
Skin symptoms	Agriculture, forestry, transport, communications	Professional & technical
General symptoms	Sales, agriculture, forestry	Service, health, hospital, social work

2.3 ECONOMIC IMPACT

2.3.1 Introduction

Although neither life-threatening nor chronically disabling, SBS is clearly perceived to be important to those affected by it. Direct evidence is limited but a significant proportion of office workers appear to be affected and its importance may increase with any movement from 'blue collar' to 'white collar' employment in the future. Extreme cases leading to building closure or demolition are extremely rare, but the more likely consequences of SBS are reduced work performance and increased absenteeism.

Because SBS may become associated with energy conservation measures it could, unjustly, militate against their wider implementation. There might be short-term value in increasing ventilation (and increasing therefore energy costs) to benefit from improved staff productivity (decreased staff costs). In practice there are two further issues to consider. The first is the environmental impact of increased energy costs: loss of non-renewable energy resources and increased pollution (contributing to climate warming and acid rain). This may in turn affect the attractiveness of the building to some organisations and individuals. The second issue is the possibility of employing better long-term solutions to SBS, which is discussed later.

The direct economic cost of SBS would be composed of reduced staff efficiency while the staff are actually working; increased probability of unofficially extended breaks and reduced overtime; increased sickness absence; increased staff turnover and increased time complaining and dealing with complaints. Although the direct evidence for these effects is limited at present, some recent studies give an indication of the scale of the problems.

2.3.2 Productivity

A re-analysis (Raw et al 1990) of data from the 'Office Environment Survey' (Wilson & Hedge 1987) indicates a strong linear relationship between the number of symptoms reported and productivity. Productivity was measured by workers' ratings of the extent to which the indoor environment affected their productivity. This measure of productivity can be regarded in two ways. First, it can be seen as a measure of what the respondent believes, regardless of whether that belief is correct. If workers believe the office environment affects their productivity, that belief is important whether it is correct or not: the belief itself may affect productivity, or the worker may leave for a job which offers a better perceived environment, and the belief is likely to affect other aspects of working life. Second, the rating may reflect actual productivity. While the scale has face validity in this respect, there is no means of establishing its actual validity since actual productivity was not assessed.

The safest assumption is that the rating is valid as a relative scale, but the actual percentage effects reported may be subject to error. This would imply that the most valid point on the scale is the zero point (zero effect on productivity) and this is the point about which conclusions about absolute level of productivity can most safely be drawn. The zero point was at two symptoms, i.e. building-related symptoms negatively affected productivity when they averaged more than two per person over a 12 month period. This result provides a valuable benchmark: more than two symptoms means a negative effect on productivity. The best buildings in the survey (none of them air conditioned) did have fewer than 2 symptoms per worker on average; the best air conditioned buildings had between 2 and 3 symptoms.

Hall et al (1991) measured the impact of SBS by questions on how frequently symptoms (a) reduced the respondent's ability to work and (b) caused the respondent to stay at home or leave work early. The response scales were not precise or linear, and 12 month recall was required. The data were also weakened by using only a binomial score (none vs one or more of 8 mucosal symptoms: dry, itching or tearing eyes; burning eyes; stuffy nose/sinus congestion; runny nose; dry throat; hoarseness; sore throat; sneezing; cough). However some useful data were obtained.

Over a third of the respondents reported reduced ability to work 'sometimes', 8% 'often' or 'always'. 3% said the symptoms caused them to leave work/stay home. A canonical correlation based on the 8 symptoms and the two response scores showed 18% common variance. Other predictors of productivity were gender, allergy, 'sensitivity', flu or chest illness, perception of hot/stuffy and dusty conditions, low job satisfaction, high role conflict and professional job grade relative to managers and administrative support staff.

Zyla-Wisensale & Stolwijk (1990) conducted one of the few studies of productivity which have used an objective measure, in this case the rate of data entry. This was not a study of SBS as such, because symptoms were not reported, but little evidence was produced of any correlation between productivity and environmental variables.

Improvement of the office environment has been found to result in higher productivity (Dressel & Francis 1987), but it is not clear whether SBS was involved.

2.3.3 Performance

Various papers have discussed the effects of exposure to toxic substances on behaviour (Colligan 1981, Evans & Jacobs 1981) and mental fatigue is one of the symptoms of SBS, but research has so far shown no measurable link between SBS and performance tests. This may be because there is no link, but this cannot be definitely concluded because of shortcomings in the studies.

Sterling & Sterling (1983a) used 'tremor' and 'T crossing' tests and failed to show a difference between the test and control groups. However the experimental and control groups were very small, only nominally matched, and clearly knew whether they were in a 'sick' or 'healthy' building.

Berglund et al (1987) tested volunteers in 'sick' and control buildings for stress, memory, vigilance, reaction time and steadiness. In order to avoid bias due to knowledge about the building, the subjects were not the usual occupants of the buildings and were unaffected by SBS. This however meant that their exposure to the building was very brief. The subjects were paid volunteers which may have masked any short term effects on the simple tasks used. Although none of the tests produced a significantly worse performance for the sick building condition than for the control building condition, trends of this expected result were found for the hand-steadiness task. This was the last task to be performed and the trend may therefore be a result of longer exposure to the building or of fatigue.

In both the Sterling & Sterling and the Berglund et al study, the tests used were relatively simple and, in the short term testing situation, liable to be affected by motivation. It is quite possible that those who were suffering most from SBS saw the tests as tests of personal ability to work effectively, rather than a test of the building. Under these circumstances they would have been more motivated to do well.

2.3.4 Absenteeism

Data on the relation between SBS cases and sickness absence has been difficult to obtain or interpret.

Broder et al (1990) carried out a simultaneous questionnaire and environment survey among 179 workers in 3 buildings. The symptoms included were eye, nose, throat and skin irritation, headache and lethargy, difficulty in concentration, and also problems with breathing, gastrointestinal problems, cough, sputum and dizziness, which are not generally regarded as SBS symptoms. Given this extended list of symptoms, it is not surprising that the number of symptoms was correlated with days absence.

In a study by Preller et al (1990), 7000 office workers in 61 Dutch office buildings reported SBS symptoms and their sick leave resulting from symptoms of any kind or leave resulting specifically from SBS symptoms. The latter classification of absence could be biased by people attributing absence to SBS because it provided an obvious justification. However, both types of sick leave were related to a range of building and personal factors, although these only partially overlapped with those factors normally considered to be correlated with SBS. Unfortunately the analysis did not examine whether there was any correlation between the number of symptoms reported and the number of days or occasions of absence.

There is more direct evidence of an effect on sick leave, although only when SBS reaches a high level in an individual worker (Robertson et al 1990). In order to control for extraneous factors, the study compared sickness absence before and after a change of workplace, thereby using each worker as his/her own control. Sickness absence was measured by number of spells and number of days, taken from actual staff absence records, and categorised by reason for absence, depending on the likelihood that the reason for absence could be directly related to SBS. However, absence for any reason could be indirectly related since uncertified sick leave may be more likely to be extended if the workplace is less attractive to return to.

Although there was a clear change in symptoms when changing building (see 3.2.3), effects on sickness absence were not clear cut. There was an overall trend for a slight fall on moving to naturally ventilated buildings (in which symptoms were less prevalent) and a slight rise on moving to air-conditioned buildings (in which symptoms were more prevalent), but the change was over all diagnostic groups. Grouping all workers in each type of building both before and after a move, a slightly higher level of sickness absence was found in those who worked in air-conditioned buildings.

The reason for the small effects on sickness absence may be found in the relation between number of symptoms and sickness absence. Combining buildings before and after the move, there was little consistent change in sickness absence until workers reported four or more symptoms at least 'most weeks'. Thus, sickness absence would only be expected to change with change of workplace if this score were exceeded either before or after the move. The sample of workers who meet this criterion was probably too small to have any realistic expectation of showing a significant effect.

The combined effect of SBS on productivity, absence from work and staff turnover is likely to have a considerable economic impact. There are no definitive UK national figures for the costs, or even for the number of buildings which could be classified as sick. BRE has performed calculations for the cost in one particular building in which the 2500 occupants reported SBS symptoms at well above average rates (evidence submitted to the UK Parliamentary Select Committee (Environment), 1991).

The calculations are based on estimates of 1 day per person per year being lost (through reduced productivity and increased official and unofficial time lost) and one hour per month per person being used for making or dealing with complaints. This amounts to a little over 1% of staff time. No allowance was made for staff turnover since there was no information available on the turnover which could be attributed to SBS. At 1990 rates of pay, the cost came to approximately £400,000 for the one building. This particular building has considerably above the average level of SBS and there is in any case considerable scope for error in generalising from one building. However, it is easy to see that the cost could run into millions of pounds annually for the UK.

In the USA, Axelrad (1989) has addressed three major types of economic costs: material and equipment damages, direct medical costs and lost productivity. A qualitative estimate of the economic costs from medical visits, sick days lost and productivity losses of white collar workers due to indoor air pollution was of the order of half a billion dollars per year whilst the estimated national annual cost of productivity losses associated with major illnesses caused by indoor air pollution ranged from \$4.7 billion to \$5.4 billion for new cases annually. He also suggested that productivity losses may be of the order of tens of billions of dollars per year. The cost estimates presented are said to be subject to great uncertainty and to be incomplete. The effects of indoor air pollution on indoor materials are a separate issue which is discussed in some detail by the US Environmental Protection Agency (EPA 1987). Damage may include corrosion of electronic components and electrical current leakage which may eventually result in equipment malfunction.

3. THE POSSIBLE CAUSES

3.1 INTRODUCTION

3.1.1 Summary

It is possible to delimit the possible causes of SBS by an examination of the mechanisms which are known to be associated with the causation of the symptoms, but the field of possible causes is still very wide. The pattern of occurrence of symptoms during the course of the working day and week, and the slow increase in symptoms when a person starts work in the building, effectively rule out an infection as the mechanism but a toxic, irritant or allergic mechanism could be associated with airborne or other contaminants. A contribution from the physical environment (e.g. temperature, humidity, lighting) must also be considered. In fact a very wide range of possible causes has been suggested: inadequate fresh air, chemical pollutants originating in the building or spread by recirculation of air, micro-organisms, noise, artificial lighting, static electricity, airborne ions, electromagnetic fields and psychological effects.

In all probability there is a different combination of causes in different buildings. Early cross-sectional studies identified a number of 'risk factors' such as air-conditioning, low quality 1970s constructions and systems, routine clerical work, open plan offices and low perceived control over the environment. The many confounders and unmeasured variables in these studies meant that it was not possible to treat these risk factors as direct causes: they could be merely statistically or causally associated with the direct cause. Only in the past few years has research begun to explore causes in controlled field experiments which have varied a range of environmental conditions to determine whether the condition varied actually contributes to the symptoms. There continue to be faults in the design, method or analysis of studies which make it difficult to draw conclusions or which have led to wrong conclusions.

3.1.2 Possible mechanisms

An examination of the symptoms and the possible mechanisms of causation provides some clues about the possible causes. This is done in more detail in the Sections headed "The potential problems" in the following text. The present Section serves only to set the scene and to delimit the possible causal factors.

Robertson et al (1985) classified symptoms as "dry" (stuffy nose, dry throat, dry skin), "allergic" (runny or itchy nose watering or dry/itching eyes), "asthma" (chest tightness/difficulty in breathing) and symptoms having an unclear cause (lethargy, loss of concentration, headache). This typology does not necessarily imply the actual cause, only the cause typically associated with the particular symptoms, but does give an indication of the range of mechanisms which might be involved.

Taking a similar approach, Jaakkola et al (1990a), in an experimental study of the effects of the use of air recirculation, defined three outcomes (mucosal irritation, allergic reaction and asthmatic reaction), based on hypothesised potential mechanisms.

Headaches can have many causes but in cases of SBS they cannot commonly be attributed to migraine (Burge 1990). Some of the symptoms of SBS would occur during infections such as colds and 'flu but in cases of SBS they typically occur more frequently and with considerably less seasonal variation than could be explained in this way.

If airborne contaminants are responsible for the symptoms of SBS, this could in theory be attributed to toxic effects, irritation, infection or immunological mechanisms.

Levels of contamination tend to be low, and often very low indeed, in relation to the appropriate occupational exposure limits. However, whilst there is as yet no proven link between SBS and the toxic effects of airborne pollutants, the toxicology on which exposure limits are based is not necessarily relevant to the low level multiple exposures in office buildings, so a toxic effect of airborne pollutants cannot be discounted.

Several reports attribute SBS, or some of its symptoms, to the irritant effect of airborne contamination. Some of the contaminants are known to have irritant properties, formaldehyde for example. Franck (1986) specifically attributes eye symptoms to drying of the eyes, possibly caused by the altered stability and composition of the eye film due to irritants such as formaldehyde, although he does not report levels of airborne contaminants. Mølhave (1990) suggests that volatile organic compounds are responsible for mucous membrane irritation, and suggests that there should be investigations to establish a dose-response relationship for organic compounds in 'sick' buildings.

Although micro-organisms may be a significant factor, SBS is very unlikely to be an infectious illness. The symptoms tend to appear at the start of the working week, increase through the week and disappear when the person is away from the place of work. Also symptoms are more frequent in the afternoon than the morning. This is not the expected pattern for an infectious illness and there has been no evidence on the symptoms being passed on outside the place of work.

The nature and time pattern of the symptoms suggests that allergenic reaction to airborne pollutants, whether chemical or microbiological agents, is possible. The observation (Burge 1990) that symptoms can take 6 to 12 months to reach maximum intensity after a person has moved into a building is also consistent with an allergic reaction. The allergenic effect of many agents is well recorded and sensitised people can be affected by even minute quantities of some agents, which might explain why symptoms occur even when air sampling fails to reveal significant levels of contamination.

Allergic reactions would however not normally be observed in such a large number of people or with such a variety of symptom patterns between affected individuals. Also, only some people who report SBS have been shown to be allergic or hypersensitive. A combination of several allergens, together with some irritant or toxic effects might account for the symptoms. The differences between air conditioned and naturally ventilated buildings can be explained if the agent originates in the air conditioning system itself or if the system creates a suitable environment for production of allergens within the building.

3.1.3 Problems with research

Many possible causes of SBS have been suggested and researched. Most explanations have focused on the air quality in the building and the systems used to ventilate the building. SBS has for example been blamed on inadequate supply of outside air, pollutants released in the building or distributed by recirculation of air, and micro-organisms breeding in humidifiers or furnishings. Other factors which have been implicated are noise, artificial lighting, static electricity, ions, electromagnetic fields and psychological effects.

Current evidence suggests that no single factor can account for SBS. Taken individually, there is evidence against (or at least no strong evidence for) each proposed explanation in at least one building or class of buildings. Different symptoms may occur in different affected individuals. One survey (Wilson & Hedge 1987) for example found lethargy to be the most frequently reported symptom, followed by stuffy nose, dry throat and headache. Other symptoms were reported at less than half the rate of lethargy. Thus, the cause of SBS may not be unitary: different components of SBS may have a different cause and the same symptom may have a different cause in different buildings.

Reports of SBS fall into two broad categories - investigation of outbreaks in order to prescribe a remedy, and more scientifically based studies to assess the extent of the problem and its likely causes. The history of SBS research studies can be seen to have passed through three phases. The first phase of research effectively commenced in the late 1970s and demonstrated that there is a real phenomenon, which we now call SBS.

Second, primarily in the mid 1980s, there were many investigations which relied on comparisons between buildings. These studies have provided evidence on what can be termed 'risk factors' - factors which are correlated with SBS but which could not be shown to be causes because of the nature of the studies. For example studies in the UK (Finnegan et al 1984, Wilson & Hedge 1987, Pickering et al 1984, Robertson et al 1985) show a number of factors which are associated with increased symptom prevalence, such as:

- air conditioned buildings;
- clerical work (relative to managerial and professional work staff);
- public sector buildings;
- open plan offices;
- low perceived control over the indoor environment.

In general, cheaply constructed buildings with poorer air conditioning systems, especially public sector buildings constructed in the 1970s, showed more problems than well constructed buildings with more expensive air conditioning systems dating from the 1980s. The type of glazing also appeared to be significant - all of the least healthy, but none of the healthiest buildings, had tinted glazing.

These factors cannot automatically be regarded as direct causes because of the many confounding factors: it is necessary to discern why each factor is significant. Confounding refers to "a situation in which the effects of two processes are not separated. The distortion of the apparent effect of an exposure on risk brought about by the association with other factors that can influence the outcome" (Last 1983). The way of dealing with confounders is to measure them or to control for them in the study design or statistical treatment.

Even with statistical controls, some problems remain. Unmeasured confounding variables may be left uncontrolled or some other model may fit the observed data equally well (for example, the researcher may be faced with the dilemma of choosing between indirect causation and spuriousness). As an example, air conditioning could be a marker for a wider range of building characteristics such as age, size, location and furnishing materials. Then the effect of age of building may mean, for example, either that buildings or services are improving in standard or that over time they are allowed to degrade and their performance therefore deteriorates.

Similarly 'public sector buildings' should not be regarded as a direct cause of SBS, but the difference between public and private sector buildings may be an important clue to causes. Wilson & Hedge (1987) suggest six possible reasons for the difference:

- quality of the buildings and building services;
- open plan offices;
- quality of building maintenance;
- susceptibility to symptoms and/or readiness to complain;
- type of work carried out;
- sampling bias.

The 'public sector buildings' did in general appear to be of poorer quality, and two case studies support this hypothesis. In one building, occupied by both public sector and private sector workers, there was no difference in symptoms between the two groups. In another case, two adjacent buildings having similar public sector workforces were distinguished by one being deep and air conditioned, the other being naturally lit and ventilated. Symptoms were more frequent in the air conditioned building. Public sector buildings also had a higher number of workers per office, more women and more clerical workers, which could have contributed to the higher symptom rates.

In addition to the problem of confounders, SBS does not occur in all buildings with the features identified by Wilson and Hedge and there were significant differences even amongst buildings having the above characteristics.

Now, in what can be seen as the third phase of SBS research, the 'risk factors' constitute important clues as to the causes, clues which are being followed up in the UK and elsewhere by making experimental changes to buildings. The basic plan of such studies is first to apply theoretical knowledge and an examination of a building to generate hypotheses about causes of SBS in the particular building being studied. Modifications are then made to the building (and thus to the indoor environment) with concurrent monitoring of both the indoor environment and the reactions of the occupants in order to determine whether the modifications have been successful in reducing the symptoms experienced. The questionnaire surveys should be carried out double-blind, i.e. neither the occupants nor those responsible for analysing the questionnaire data should be informed what modifications are being carried out, or when.

This approach controls for most confounders and provides much stronger evidence for the environmental causes of SBS, because most psychological factors and environmental factors remain unchanged. Including a wide range of possible determinants in a study also permits the researcher to identify interactions among factors and different causes in different cases. In a statistical analysis of several buildings, there is a danger of diluting significant effects in individual buildings if only single potential determinants are examined. In the discussion which follows, multifactorial experimental studies are therefore given greater weight in assessing the evidence on causes of SBS.

All three approaches have validity for specific purposes, but each needs to be done well if meaningful results are to be achieved. One notable difficulty is that many investigations are limited in approach so that they often consider the medical aspects, or measurement of airborne contaminants or an assessment of the physical environment and the ventilation or air conditioning system, but not all three aspects simultaneously. This means that some investigations have drawn conclusions which do not appear to be fully supported by the reported facts, possibly because not all of the information found in the investigation is reported or because the investigators have made assumptions. Whatever the reason, some conjecture appears to be subsequently reported as fact, especially in the press. Unfortunately, few investigations have conclusively proven a cause by the successful application of remedial measures.

It is useful to explore the reasons for null results in certain studies, and a study by Nelson et al (1991) illustrates a number of pitfalls. This was a survey of environmental conditions and user evaluations in the three headquarters buildings of the US Environmental Protection Agency. An initial staff survey was used to select monitoring sites to maximise variability in the indoor conditions, and data collection carried out daily at the sites selected.

The questionnaire had 33 health items, which were reduced to 11 by principal components analysis. This resulted in clusters with a mix of SBS and non-SBS symptoms, and different clusters to other studies which used the same approach (Hall et al 1991) and even the same buildings (Wallace et al 1991). This is in general not good practice when (if the researchers thought it necessary to have clusters at al) the clusters could have been generated from larger and more representative previous studies.

All data were converted to binary codes, thus losing data. The variables were classified as independent variables (the results of environmental monitoring) and confounding variables (items from the questionnaire or observations by the researchers), which could alternatively have been regarded as independent variables (e.g. hours spent at a VDU, use of chemicals).

Using data from the final sample of respondents (less than 200), 66 separate statistical models were developed using logistical regression analysis: 3 models for each gender and symptom cluster. There were no consistent patterns of results, but this cannot be taken to mean that none of the factors measured had an effect on SBS because of the many problems with the method.

As another example, Broder et al (1990) carried out a simultaneous questionnaire and environment survey among 179 workers in 3 buildings. The symptoms included problems with breathing, gastrointestinal problems, cough, sputum and dizziness, which should not be regarded as SBS symptoms. Work-related symptoms were correlated with the individual's perception of the environment but not with the actual measured levels. It is not clear from the report of the analysis whether objective environmental variables might have entered the equations had the perceived variables not already been there. It is also not clear whether the objective monitoring did adequately characterise the environment, and the response rate was only 20%.

There has also been some danger of drawing conclusions from reports of problems identified in the indoor environment without linking the problems to symptoms or complaints. For example Turner & Binnie (1990) report a wide-ranging survey of indoor air quality without reference to occupant health or evaluation of the environment. This is all useful data but in reality indoor air quality can only be defined by effects on occupants, whether the effects are inferred from current air quality standards or measured in the same context as the pollutant level measurements.

An American study of 4 buildings resulted in 7 papers (Crandall et al 1990, Fidler et al 1990, Hurrell et al 1990, Leaderer et al 1990, Nelson et al 1990, Persily & Dols 1990, Selfridge et al 1990) which report a wide range of contaminants in addition to symptoms, complaints about the environment and sources of job-related stress, but there was (initially at least) no correlational analysis. This kind of analysis of 4 buildings is of little value - larger surveys are able to provide normative statistics for comparison with subsequently surveyed buildings.

3.1.4 Presentation of the evidence

The following Sections of this report assess the research evidence on the range of suggested causes of SBS in more detail, dealing with the following issues:

- ventilation (fresh air rates and type of system);
- indoor air pollutants (tobacco smoke, simple inorganic gases, volatile organic compounds, micro-organisms, non-viable particulates and fibres);
- hygrothermal factors (temperature, humidity and air movement);
- lighting;
- noise and infra-sound;
- electromagnetic factors;
- psychological factors and individual characteristics;
- management and organisational factors.

3.2 VENTILATION

3.2.1 Summary

The role of ventilation in SBS has been debated in two different but related terms: the ventilation system and the supply of fresh air.

The lowest mean symptom prevalences in the UK are found in naturally ventilated buildings and mechanically ventilated buildings. In Scandinavia there is probably some excess risk from mechanical ventilation, possibly because this decreases indoor humidity. Mean levels are clearly higher where there is cooling capacity in the ventilation system, but only a small additional risk appears to be present where there is humidification.

Partly because of the association between SBS and air-conditioning, inadequate ventilation has been high on the list of causes proposed by many commentators, but this has been largely due to the many reports of problems with ventilation systems in 'sick' buildings. These reports do not in the main result from controlled trials or correlational studies, and many originate from North America at a time when typical fresh air rates were relatively low because of the energy crisis of the mid 1970s. There are many reasons why ventilation might be inadequate in buildings: ventilation systems which are malfunctioning, badly maintained, badly designed or ill-controlled can reduce the amount of air supplied to a level below a specified minimum and the distribution of fresh air within the occupied space may be inadequate. More simply, windows may be kept closed to maintain comfortable temperature conditions.

In spite of all these potential problems, which are too often realised in practice, the evidence suggests that in practice low ventilation rates are not a major determinant of SBS. Evidence for a correlation between ventilation rate and SBS symptoms is patchy at best, air-conditioned buildings generally have higher ventilation rates than naturally ventilated buildings and the few controlled trials which have been carried out have mostly failed to find an effect of varying ventilation rates over a very wide range.

While very low ventilation rates will obviously cause problems, and increasing ventilation will reduce pollutant levels, current ventilation standards appear to be appropriate on the whole (assuming the outdoor air is not highly polluted) and there is little advantage to be gained from increasing them further unless there is no alternative means of reducing pollution in a particular building. Increased ventilation has a cost in economic and environmental terms, requiring energy and sometimes causing new problems such as draughts.

The most crucial aspects of ventilation to be addressed in current practice are probably the maintenance of ventilation systems (which has a direct impact on levels of indoor contaminants) and the distribution of air within the occupied space.

3.2.2 The potential problems

Of all the typical features of 'sick buildings', the system of ventilation is often regarded as the most significant. This leads to a presumption that lack of fresh air is a major cause of SBS. In reality the ventilation system and the ventilation rate need to be considered as different, although related, factors.

'Fresh air' is required for various reasons, the main ones being to supply air for respiration, to dilute contaminants and to reduce odours. Ventilation, although not necessarily fresh air, may also be required to maintain personal comfort, i.e. for the control of air temperature and humidity. These factors are considered in Section 3.4.

Several standards have been set for ventilation and fresh air supply rates to offices. In the USA, the American Society of Heating, Refrigeration and Air Conditioning (ASHRAE) standard is most widely accepted and has been periodically revised, as have the standards set by the Nordic Committee on Building Regulations (NKB). Studies by Yaglou et al (1936) resulted in a ventilation rate standards of 7.5 L/s/person (ASHRAE 1977), reduced as a result of the energy crisis, for example to 2.5 where smoking is not permitted (ASHRAE 1981) or 4 L/s (NKB 1981).

These two standards have now been raised again to 10 (ASHRAE 1989) and 11 L/s (NKB 1991). The current ASHRAE Standard is based on achieving an indoor CO_2 level of 1000 p.m. and 80% of the occupants should be 'satisfied'. In the UK, the Chartered Institution of Building Services Engineers (CIBSE 1991) sets a comparable standard which ranges from 8 litres per second per person in general offices with no smoking, up to 32 litres per second per person for personal offices or boardrooms where smoking is heavy.

These standards are all based primarily on people being the main polluter. Fanger (1988a,b) has conducted research to throw considerable doubt on this. Using the odour criteria applied to pollution from people suggests that an additional 40 L/s could be necessary for pollution originating from the building. However it is not clear from Fanger's work whether some of the pollution from the building actually originated from people but had been transferred to 'sinks' (pollutants deposited on surfaces) in the building and its services.

The impetus to seal buildings, increase control over the environment and reduce ventilation is usually motivated either by necessity ('deep' buildings with open plan offices are difficult to ventilate naturally) or by a desire to save energy (and money). The practice of tight control over the indoor environment poses problems if the ventilation or air conditioning system is in any way imperfect.

The practice, rarer in the UK than in the USA, of using CO_2 monitors to control ventilation may also cause problems. Janssen & Hill (1982) describe such a system set to operate at 2500 ppm CO_2, i.e. the system would not draw in fresh air until the CO_2 level reached 2500 ppm. This criterion CO_2 level was too high and, since CO_2 levels did not rise above 1800 ppm the system did not draw in fresh air at all. Such cases would now be unlikely to occur, and CO_2 monitors have improved.

Indeed it is now possible (Vaculik & Plett 1990) to have CO_2-controlled ventilation which includes an element of the CO_2 concentration which would be expected if the system were performing satisfactorily (based on concentration history). This is an improvement but the relation of CO_2 to other pollutants may be variable and CO_2 in exhaust may bear a different relation to CO_2 in the breathing zone under different weather conditions. For example in a leaky building the infiltration rate will depend on external winds and the indoor - outdoor temperature difference, and air mixing within the building will depend on surface temperatures. Another option now available is to use sensors which detect a mixture of gases (see Raatschen 1991) and the next few years will probably see refinements to this approach.

Even if the appropriate volume of fresh air is delivered to a space, it may not reach the occupants in the way intended: a certain amount of the ventilation air may be extracted first, rendering the ventilation effectiveness (and fresh air efficiency) less than 100%. Ventilation effectiveness depends on supply air temperature, type of supply diffuser, relative position of supply and return air inlets, partitioning of the office space, size of office, height of partitions and gap between partition and floor, orientation of partitioned space, workstations and supply/return inlets (Farant et al 1991). Air distribution is therefore particularly susceptible to problems caused by changes in the layout of office spaces and the unregulated covering or adjustment of air inlets by building occupants.

Lack of fresh air, stuffiness and too much air movement can occur in naturally ventilated buildings as well as in air conditioned mechanically ventilated buildings. The reasons may be different in the two cases. In the case of natural ventilation, fresh air is provided by uncontrolled infiltration through the building fabric or by controlled ventilation through windows and this is found in the majority of UK buildings. The use of windows permits a wide degree of local control of ventilation and is frequently also used for temperature control (Warren & Parkins 1984). While this appears a satisfactory solution to many occupants it is not without its problems:

- what is satisfactory ventilation for the person opening the window may be a gale to an adjacent occupant;
- the energy required to heat excessive fresh air input is wasteful;
- any control system used in the building systems design is unlikely to be

capable of controlling the indoor environment - however good the design.

In air-conditioned buildings the design will normally be based on the assumption that the windows will remain closed and the building services should therefore provide the correct quantities of fresh air, proper distribution and circulation of air, while avoiding draughts. The selection of adequate fresh air quantities is reasonably straightforward in principle but there can be complications because the amount of air in circulation, needed to carry heating or cooling power, often exceeds the fresh air requirements. This discrepancy may be overcome in two ways: either use more than the minimum fresh air requirement or use the minimum and mix it with air recirculated from the occupied spaces to make up to total quantity.

In practice there is frequently a compromise, where the total quantity is kept constant but the ratio of fresh and recirculated air is varied according to the heating and cooling demand. For example in mid-winter the minimum fresh air quantity is used with maximum recirculation to use the recirculated air for heating purposes. In milder weather less heat is required so the fresh air quantity is increased and the recirculation air is reduced, and so on. This is accomplished by a control system through motorised dampers in the air ducts. The correct sizing of the ducts and fans can cause problems and so can imprecise specification of the control system and/or poor commissioning.

There is a further complication on variable air volume (VAV) systems which can affect fresh air quantities. When a VAV system is operating at its design volume the operation of mixing fresh air and recirculated air is the same as just described but difficulties arise when the total volume of air through the plant is reduced. For the sake of simplicity assume that the minimum fresh air quantity is 25% of the total volume at a particular time and that to achieve this the dampers on the fresh air intake are 25% open. If the VAV system now turns down to only 50% of the initial total volume then the fresh air quantity will only be 25% of the 50%, i.e. 12½% of design volume. To compensate, the minimum fresh air damper is moved to the 50% open position giving a fresh air quantity of 50% of 50%, i.e. the equivalent of 25% of the initial total. Unless this feature is both specified and adequately controlled it is quite likely that the minimum fresh air quantity will not always be achieved. Incomplete commissioning and poor maintenance will exacerbate this problem.

The effect of filters on fan sizing is also important as a dirty filter will reduce the volume of air as compared to operation in the clean condition.

Mechanical ventilation of buildings therefore differs from natural ventilation in a number of other ways, most significantly:

- whilst mechanical ventilation and air conditioning can exercise more precise overall environmental control they allow less personal choice or local control;
- mechanical ventilation and air conditioning systems have components that are susceptible to failure and to poor specification, design, installation, commissioning and maintenance;
- ventilation systems can harbour organic growth (see 3.3.2) and may distribute contaminants from one area throughout the building.

People often express a wish to have openable windows, but this probably has as much to do with control over the indoor environment and contact with the outside world as it has to do with the amount of outside air entering the building.

A background level of ventilation, based on the number of occupants, is always necessary, but any additional ventilation is determined by the likely indoor pollutants and cooling requirement. It also needs to be recognised that increasing ventilation can introduce new indoor environment problems, for example draughts, increased transfer of contaminants from the ventilation system and increased disturbance of settled dust in the building.

3.2.3 Evidence for effects on SBS

It is often said that inadequate ventilation is a common denominator in cases of SBS. Inadequate ventilation was identified as a suspected causal factor in 50% of 356 investigations conducted by NIOSH between 1974 and 1985 (Wallingford & Carpenter 1986, NIOSH 1987) and 68% of Canadian buildings investigated by Health and Welfare Canada in 1984 (Kirkbride 1985). There has been similar experience in Denmark (Valbjørn et al 1990a). Nevertheless it is far from clear that low ventilation rates are actually responsible for many cases of SBS. In spite of the weight of circumstantial evidence, the fact that two phenomena are observed in the same building does not provide sufficient evidence that the one causes the other. In any case these investigations were conducted prior to the increases in most ventilation standards over the past two years.

Tales of dissatisfaction amongst building occupants because of the ventilation system are also legion. Youle (1986) and Waller (1984) detail a number of such problem buildings where an excessive number of complaints were received because of inadequacies in the ventilation or air conditioning systems due to poor design, installation and maintenance. Since the reports give no details of medical symptoms, these cases cannot be considered as SBS but they do indicate a possible contributory factor. Nevertheless, occupants' complaints always require careful interpretation. The best policy is probably to believe occupants when they say that **something** is wrong, but to understand that they can be misled regarding **what** is wrong.

There can be little doubt that increasing the ventilation rate or reducing the proportion of recirculated air will normally reduce indoor pollution. For example Farant et al (1990) found that with higher outdoor air rates, concentrations of carbon dioxide, oxides of nitrogen and VOCs were lower. Turiel et al (1983) found that indoor concentrations of contaminants are correlated with the proportion of recirculated air, but none of a total of 28 measured contaminants exceeded recommended outdoor levels (EPA 1971, OSHA 1975) in a building which exhibited SBS.

Recirculation is not always necessary on grounds of energy efficiency since total energy recovery systems can provide an alternative (Bayer & Downing 1991).

The real issue is whether current ventilation rates are likely to cause problems. For example, Putnam et al (1990) investigated 6 buildings, 4 of them already believed to be 'sick'. CO_2 levels were all below the 1000 ppm ASHRAE 62-1989 standard, and did not relate to complaints (assessed by questionnaire), ventilation problems or other major pollutants.

Neither does a drastic change in ventilation necessarily solve problems. Landrus & Witherspoon (1990) report a study in which new openable windows were installed in a school from which complaints had been received about indoor air quality. The problems were not solved, although there was a short-term political solution to the problem. This emphasises the need to consider all aspects of the environment and not to rely on building occupants to define solutions. In another case study, Donnini et al (1990) report that increasing the amount of outdoor air reduced CO_2, formaldehyde and toluene, but increased dust, temperature and relative humidity. Questionnaire reports of IAQ complaints increased.

In a cross-sectional study, Hodgson et al (1990) found SBS to be associated with lower air flow rates, but SBS was defined by headaches, nosebleeds, eye irritation, sore throats and chest tightness. Of these only two are standards on SBS questionnaires, whereas skin symptoms, nose irritation and dry throat were excluded.

Shen (1990) found no decrease in complaints from a building when ventilation rates were increased, but initial rates complied with Chinese standards. Supplying air direct through a filter into the rooms did reduce complaints but there was no account of the complaint assessment method. The introduction of new filter units would have been more obvious to the occupants than the change of ventilation rate and this may account for the change in complaints.

Sundell et al (1991) studied three groups of symptoms: general, mucous membrane and skin symptoms, based on those symptoms which were reported often (every week). Building-relatedness not established. The questionnaire survey preceded environmental measurements and therefore referred to different times and seasons. Four types of building were studied (mean outdoor air supply rate in l/s per person in brackets): naturally ventilated (4.6) and mechanically ventilated with exhaust only (8.6), supply and exhaust (20.0) or supply and exhaust with heat exchanger (21.1). There were no air-conditioned buildings in the sample; these are not common in Sweden.

Symptoms (particularly general symptoms) were lower in buildings with higher outdoor air supply, 'sensation of noise' was higher, 'sensation of dry air' was lower. However other factors were not controlled except sex and three types of building design. There was very little variation in symptoms between buildings with different ventilation systems.

Berglund et al (1988) studied the effect of different recirculation rates used in autumn (no recirculation) and winter (76% recirculation) in a library. There were no differences in symptom rates between the two conditions, even though indoor pollutant levels were approximately doubled under the recirculation condition.

In an experimental study by Jaakkola et al (1990a), 75 office workers were exposed to different ventilation conditions, alternating one week in each condition. In the high ventilation condition there was zero recirculated air and 23 L/s/person of outdoor air. In the low ventilation period there was 70% recirculation, resulting in only 6 L/s/person.

There was no difference between the two conditions in reported symptoms or perception of indoor air quality, neither was there any significant difference in smoking or opening windows between the two periods. It may be that one week is too short a period for this kind of study because of the possibility of pollutants being transferred to sinks during periods of low ventilation, to be re-emitted during periods of high ventilation. A week may also be too short a period for occupants to recover from sensitisation to pollutants.

Although the gross change in ventilation had no effect, an analysis of variations in ventilation rate within ventilation conditions (Jaakkola et al 1990b) did show an effect when the ventilation rate was low: SBS symptoms increased as the ventilation rate decreased below 15 L/s/person.

Menzies et al (1991) switched the outdoor air rate in two buildings between nominal rates of 10 and 25 l/s/person, one week at each level, for 6 weeks in a double-blind cross-over design. A weekly questionnaire was used, together with monitoring of CO_2, temperature, RH, air velocity, fungi, aeroallergens of house dust mite, dust, HCHO, NO_x, TVOC, noise and light in the rooms. The first 3 of these variables were also monitored in the HVAC system and outdoors.

The CO_2 and VOC measurements indicated that the changes in outdoor airflow achieved were not the intended 10:25 (1:2.5) but 1:1.3 in one building and 1:1.8 in the other, with little variation in temperature and humidity. There was no difference in symptoms or environmental evaluation by respondents between the two conditions but symptoms and environmental ratings declined steadily over the period. Neither were symptoms correlated with CO_2 or VOC levels.

Nagda et al (1991) switched the nominal outdoor air rates from 10 to 17.5 l/s/person in two 2-week periods (monitoring only in the second week) in summer and in winter. Monitoring included RSP, air velocity, CO_2, CO, temperature, humidity, VOCs, and microbial aerosols. The actual ratios of air change rates between the high ventilation and low ventilation conditions were 1.28 and 1.61 for the summer and winter respectively. There were generally low pollutant levels and only small differences in CO_2 and other pollutants between conditions.

Comfort ratings were similar in each condition and the percentage of occupants experiencing one or more symptoms was higher under low ventilation conditions for the summer trial only. This result is difficult to interpret because the method of analysis loses data; the study was done in a single building, with no control group or cross over; symptoms were checked for 'building-relatedness' by asking if respondent thought they were related to IAQ (this is a leading question and is not the normal approach) and symptoms included asthma and fever in addition to SBS symptoms.

The role of different methods of ventilation also needs to be made clear. SBS has been found even where there is not recirculation of a proportion of air from the building. Also, it is possible to have an air-conditioned building which is as healthy as a typical naturally ventilated building (Dixon 1991) and it is also possible to have a naturally ventilated building with high levels of complaints.

A reanalysis of data from 3 British studies of SBS (Mendell 1990) enabled a clearer picture to be gained of the effect of ventilation type. The symptoms from naturally ventilated and simple mechanically ventilated buildings were very similar. Where air conditioning was present, the rate of symptoms was higher.

The addition of either steam- or water-based humidification had little, if any, effect, although there was some indication that water-based humidification would increase symptoms in some circumstances. It is likely that the effect of humidification is ambiguous because it can have different effects in different circumstances. Where it promotes the growth of micro-organisms it would have an adverse effect. Where it lifts the humidity of the air above 20 - 30% it could have a beneficial effect (see 3.4.3). It is likely that the results reported reflect a combination of these two trends.

Robertson et al (1990) investigated whether SBS symptoms are affected by changes of workplace (all office buildings). The number of symptoms reported showed a significant rise after moving from a naturally ventilated to an air-conditioned building and a significant fall after moving from an air-conditioned to a naturally ventilated building. Ventilation category is unlikely to have been the only difference between the buildings concerned. The improvement in work-related symptoms, on moving from air-conditioning to natural ventilation, was not as great as the deterioration on moving from natural ventilation to air-conditioning. It may be that the interval for recovery is long or that, once workers have SBS, even low levels of exposure to causal factors continue to give rise to symptoms. Alternatively, this asymmetry may be accounted for by the retrospective reports of symptoms in previous buildings.

In a study of day care centres (Sverdrup et al 1990) those with supply and exhaust mechanical systems had higher symptom rates than those which were naturally ventilated or had a mechanical exhaust only. The buildings with supply and exhaust systems had slightly higher background noise levels and infrasound levels, although they had similar VOC levels to those with mechanical exhaust and lower levels than those that were naturally ventilated. VOC levels followed the same pattern as the ventilation rate, which was highest in those which were mechanically ventilated.

Burge et al (1990a) carried out a paired comparison of a relatively sick and a relatively healthy building, with a pair of buildings in each of three ventilation classes - naturally ventilated buildings, air-water air conditioned buildings with induction units, and variable air volume all-air systems. Questionnaire surveys were carried out together with environmental monitoring in one area of each building. All four air conditioned buildings had far higher air change rates than the naturally ventilated buildings and there was no difference in ventilation rate between the 'sick' and the 'healthy' building in each pair.

The 'sick' naturally ventilated building was slightly warmer and drier than its healthy counterpart, but the opposite relationship held with both types of air conditioned building. All the environmental parameters were within CIBSE recommended standards. Air speeds, lighting levels and sound levels also did not discriminate between the 'sick' and 'healthy' buildings.

The 'sick' air-conditioned buildings did however show inadequate maintenance, poor documentation and records of the use of the system. While these factors may have contributed to the problems, it is not clear in what way they would, given the level of environmental monitoring that had been carried out. It seems possible that poor maintenance could, however, result in greater variability of temperature and humidity, and in higher levels of air pollutants which were not measured in this study.

Zweers et al (1990), in a major cross-sectional study, found that skin symptoms were additionally significantly affected by the presence of a spray humidifier, eye symptoms and neurological symptoms were higher where there was mechanical ventilation, nose and throat symptoms were higher where there was steam humidification.

Stanwell-Smith (1986) found an effect of proximity to air vents on tiredness, eye problems and shortness of breath, but it is not clear whether this was due to pollutants or to draughts.

Hansen (1989) studied SBS in six day-care institutions before and after removal of man-made mineral-acoustic ceilings and installation of mechanical ventilation. The intervention was associated with a reduction in symptom prevalences and a decrease of symptoms, but the subjects were aware of the changes, and this awareness could have caused part or all of the effect and there was no control group.

In summary, an inadequate volume of fresh air is unlikely to be a sufficient explanation for SBS, although it can be a contributory factor, for a number of reasons:

- ventilation rates can be, and often are, higher in air-conditioned buildings than in naturally ventilated buildings;
- a high ventilation rate does not guarantee a high fresh air rate or that fresh air is being distributed to the occupants - design of air supply and extract is critical;
- in some buildings it has proved difficult to show a change in SBS by varying the rate of fresh air supply to a building over quite a wide range (e.g. 0-70% recirculation/6-23 litres per second) and buildings with high ventilation rates (well in excess of the ASHRAE minimum) can be 'sick';
- within a ventilation category, there is no demonstrated correlation between ventilation rate and symptom prevalence.

In any case, where the cause of SBS is thought to be indoor pollution, it should be dealt with primarily by removing or isolating the source of the pollution. The ventilation rate per person is not a sufficient measure of air quality since there are many sources of pollution other than persons. Unfortunately there is not at present adequate data on how much additional ventilation might be required, since this would vary greatly from building to building.

It continues to be incumbent on building designers and operators to ensure that system commissioning, maintenance and cleaning are both feasible and actually undertaken to minimise any possible contribution of these factors to SBS.

Inadequate ventilation can only be defined in the context of comfort conditions and the rate of production of pollutants, and therefore the indoor environment is the issue and ventilation is only one aspect of it.

3.3 INDOOR AIR POLLUTANTS

3.3.1 Summary

There are many well-documented potential and actual sources of pollution in buildings, including the occupants, the buildings and furnishings, office materials and equipment, building services and the environment outside the building. There is consequently a wide range of pollutants in the indoor air. However, each pollutant is typically found at levels well below any established occupational exposure limit values and it has been difficult to show clear differences between 'sick' and 'healthy' buildings in the levels of pollutants.

Studies have not shown a higher symptom prevalence in buildings in which smoking is allowed, but there may be higher ventilation rates to compensate in such buildings, and possibly an element of self-selection of staff. Studies which have compared individual workers who are exposed or not exposed to environmental tobacco smoke have often found a higher symptom rate in the exposed population.

Simple inorganic gases such as carbon dioxide and ozone may occasionally present problems but are unlikely to be widespread causes of SBS.

Volatile organic chemicals (VOCs) such as formaldehyde are evolved by a large number of materials and processes in buildings and indeed by the occupants, and can reach moderately high levels in new buildings. Few studies have examined the role of VOCs in SBS in a controlled ways but those that have done this have tended to show that higher levels are associated with more symptoms. The measurement and effects of mixtures of large numbers of VOCs is not sufficiently clearly understood to be able to carry out definitive studies: we cannot make clear inferences about a variable which we do not fully understand how to measure. The role of VOCs in SBS is likely to be greater in newer buildings and is unlikely to be a major factor in older buildings unless there are significant sources other than the building itself (e.g. new furnishings).

There are many examples in the literature of serious contamination in ventilation systems, and clear evidence that bacteria and fungi can colonise a wide range of habitats in buildings and building services where there is the least failure of maintenance. What is less clear is whether the actual level of airborne organisms is at all correlated with symptom prevalence. There is limited positive evidence from cross-section studies and cases studies, but no good controlled trials. However most of these studies did not consider the metabolic products of micro-organisms or non-viable fragments. SBS has been more clearly shown to be related to the size of areas in which dust and dirt can gather in buildings, which provide breeding sites for micro-organisms and invertebrates such as mites, and the standard of office cleaning. Micro-organisms have a higher probability of occurring in older buildings.

Recent studies have increasingly been showing that airborne dust levels are a significant factor, although the contribution of the organic and inorganic components of the dust, and gaseous pollutants adsorbed onto it, has not been clarified. Dust cannot be regarded as simply an air pollutant, effective only when inhaled, since it can be transferred direct to the skin or ingested with food or drink; these aspects have hardly been explored at all.

3.3.2 Sources of indoor air pollution

(a) INTRODUCTION

Pollutants may be introduced into a building from a very large and diverse range of sources. The sources can be broadly categorised as:

- building users and their activities
- building and furnishing materials
- office materials and equipment
- heating, ventilating and air conditioning systems
- the outdoor air and soil.

The types of pollutants which originate from each source are also many. Those of primary concern in relation to SBS are:

- simple inorganic gases (mainly oxides)
- volatile organic chemicals
- non-viable particulates and fibres
- micro-organisms/viable particulates.

Water is, in some circumstances, rightly regarded as an indoor pollutant considered under separate headings in this report because of its dual e the direct effect of relative humidity and the indirect effect via the gr and dispersal of micro-organisms. The concentration and balance of positive and negative ions in the air is also regarded by some as being of importance, but this is dealt with under the heading of electromagnetic effects.

Even using the simplified categorisations above, it is clear that there are many combinations of sources and pollutants. In addition the sources may not be independent, for example ventilation systems may recirculate pollutants from other sources and furnishings may absorb and re-emit pollutants. The potential range of pollutants in the office and similar environments is therefore enormous. However, levels of individual pollutants actually occurring have generally been found to be minute, sometimes requiring techniques more sensitive than normal occupational hygiene practice for their measurement.

It is therefore difficult to carry out research on particular sources and pollutants while excluding the effects of others, and also far from simple to summarise the evidence. Therefore the presentation of the evidence in this review is different to that used for the other possible causes. This Section briefly summarises the types of pollutant which can originate from different sources, organised by source rather than pollutant. The following Sections deal in more detail with the potential problems and the actual evidence regarding each pollutant or pollutant group. The summary at 3.3.1 above draws conclusions about the evidence regarding indoor air pollution as a whole.

(b) BUILDING USERS AND THEIR ACTIVITIES

Pollutants released by the occupants themselves include CO_2, water vapour, microbial organisms, organic vapours and particulates such as skin scales.

In addition the personal activities of the occupants (as distinct from work activities such as the use of chemicals or equipment) can influence indoor pollution levels. The most obvious example of this is probably smoking, but the use of cosmetics, eating and drinking in the workplace all have a potential contribution to total pollution levels.

Occupants are also important for the ways in which they may attempt to improve the indoor environment, for example by seeking to modify ventilation openings or introducing green plants to the office, but there is little evidence about the influence that occupants have in this way.

(c) BUILDING AND FURNISHING MATERIALS

Many sources of pollution have been attributed to emissions (or 'off-gassing') from the fabric and furnishings of the building. The primary cause for concern has been VOCs emitted from new building materials and furnishings such as foams or particleboard. Other new materials can also release pollutants or exacerbate problems by the release of water vapour. Older materials can cause different problems if the materials themselves decay or if they become a reservoir of dust and dirt, particularly organic material which causes pollution itself or which provides a breeding site for micro-organisms.

(d) OFFICE MATERIALS AND EQUIPMENT

Chemicals are used in offices for a number of purposes, including cleaning furnishings and correcting text. Office materials such as paper (particularly rinted glossy paper) and plastics can also contribute to the pollution load. The chemicals from these sources would be mainly VOCs.

Photocopiers have also been suggested as a cause of SBS because pollutants such as ozone and organic vapours can collect in inadequately ventilated rooms (Rajhans 1983).

(e) HEATING, VENTILATING AND AIR CONDITIONING SYSTEMS

Heating, ventilating and air conditioning (HVAC) systems can contribute to indoor pollution in three main ways: providing insufficient ventilation air (see Section 3.2), transmitting pollutants from one part of a building to another and emitting or producing pollutants from within the system itself.

The transmission of pollutants from one part of a building to another can obviously apply in principle to any pollutant present in the building. The probability of transmission is greater for gaseous pollutants than for particulates, assuming the air filters in the systems are appropriate and in good order. However aerobiologists in the health service have shown that pathogens can be carried in hospital ventilation systems, thus spreading infection (Sykes 1989). This indicates the potential for viable particulates and even large non-viable particulates and fibres to be transmitted. Such problems will be made worse by malfunctioning ventilation systems and by the problems inherent in achieving good mixing of fresh and recirculated air (MacDonald 1991).

HVAC systems themselves can emit pollutants, for example VOCs from new filters or ductwork, fibres from filters or duct lining (e.g. Pejtersen et al 1989). Heating coils can also collect dust and dirt and subsequently produce odours. Probably more importantly, HVAC systems can provide a site which micro-organisms can colonise and in which they can proliferate. Cold-water humidifiers have been the focus of interest here but other items of plant can also act as breeding sites. For example, air conditioning cooling coils where condensed water can collect to produce a covering of slime have been shown to release micro-organisms into the airstream (WHO 1979).

Unflued heaters and flues close to ventilation intake are able to release toxic gases - particularly carbon monoxide and oxides of nitrogen. Where heating elements are present in the office space, combustion of dust can result in low levels of pollution.

(f) THE OUTDOOR AIR, WATER AND SOIL

The outdoor air has usually been regarded as 'fresh' air for the purposes of ventilation design. This position has now changed and ASHRAE (1989) standards now require that the air used for ventilation must be filtered to remove outdoor pollution unless it already meets US Federal guidelines. In practice this means that opening windows are not an acceptable means of ventilation in highly polluted areas. The contribution of the ambient outdoor air cannot therefore be ignored in considering the causes of SBS, although in most cases there is no evidence of the outdoor air being sufficiently polluted to cause problems.

The siting of ventilation intake close to localised pollution sources (e.g. car parking, ventilation exhaust) is another matter, and has been a problem in some buildings. Perhaps more important from the point of view of maintenance staff is that a high rate of ventilation with polluted air will probably result in any symptoms experienced by occupants being blamed on the system.

The main pollutants to enter buildings from the soil are radon and gas from landfills and other contaminated land. Radon is associated with long-term effects on lung cancer but there is no reason to associate it with SBS. Landfill gas contains methane, carbon dioxide and other gases in lower concentrations, but has not to date been associated with any cases of SBS: the primary hazards are explosion and asphyxiation.

3.3.3 Environmental tobacco smoke (ETS)

(a) THE POTENTIAL PROBLEMS

Smoking is an obvious source of airborne contamination. Hobbs et al (1956) and Brundrett (1975) list some of the many gases to be found in cigarette smoke, which include carbon monoxide, carbon dioxide, nicotine, aldehydes, ammonia, phenols, hydrogen cyanide, pyridines, oxides of nitrogen, and acrolein. There is also a significant particulate component in ETS. Many of these components have toxic or irritant properties; some are considered separately under other headings in this review, but tobacco smoke was felt to be a sufficiently specific, important and contentious issue to merit a Section of its own.

Some researchers have suggested that smoking does not in practice affect contaminant levels: Sterling et al (1987), in reviews of NIOSH and CDSC reports of 350 investigations, found lower levels where smoking was allowed than where it was not allowed. However, this report does not discuss the building ventilation rates and may therefore not give an accurate picture. Since 1981, the American Society of Heating, Refrigeration and Air Conditioning Engineers (ASHRAE 1981) has advocated a higher ventilation rate where smoking is permitted (7.5 l/s/person) than where it is not permitted (2.5 l/s/person) and this may well account for these anomalies. Other research by the US Environmental Protection Agency (Wallace & Bromberg 1984) identified higher levels of contamination in residential buildings where smoking was permitted than where it was not permited.

In addition to direct exposure to ETS, there can be substantial secondary exposure due to adsorption and re-emission from building materials. Iwata et al (1990) showed that tobacco combustion products, detectable as odour, readily adhere and re-emit from many surfaces, even aluminium foil.

Several studies (Pimm et al 1978, Shephard et al 1979a,b) have shown exposure to cigarette smoke to induce eye irritation, nose irritation, stuffy or runny nose, shortness of breath, sore or dry throat, cough, tightness in the chest, wheezing, dizziness and nausea. Some of these symptoms are in common with SBS. Brundrett (1975) also lists physical symptoms caused to smokers and passive smokers which overlap with symptoms of SBS.

(b) EVIDENCE FOR EFFECTS ON SBS

Evidence on the role of ETS is ambiguous. Those studies which have focused on the exposure of the individual person have tended to show an effect of ETS. This was the case in the major British study (Wilson & Hedge 1987) and in studies in Finland (Jaakkola et al 1990b). A more detailed analysis of the British data (Robertson et al 1988) suggests that ETS may be a more significant factor in naturally ventilated buildings, perhaps because other factors which contribute to SBS are less evident.

A major Dutch survey (Zweers et al 1990) found nose and throat symptoms were higher among smokers but the analysis did not detect a significant effect of passive exposure to ETS. Urch et al (1990) showed that nasal symptoms and headache, although not the other symptoms of SBS, are correlated with exposure to smoke in non-asthmatics. In asthmatics there was additionally a correlation with eye and throat symptoms and fatigue.

In contrast, studies which have compared buildings in which smoking is or is not allowed have generally shown null results.

Hedge et al (1991a) compared SBS in buildings with a range of smoking policies. Pollutant levels were higher in smoking areas but SBS was not related to smoking policy. Similar results were found in several previous studies (Sterling et al 1983b,1987, Hedge 1984a,b) and a study of a single building (Taylor et al 1984). The null results may have been due to a number of factors. There was some variation among smoking categories in company, type of location, age, size, HVAC system, design air change rate, presence and type of humidifier, environmental monitoring strategy and questionnaire return rate. There was no attempt to control for any of these factors and no comparison of symptoms between smoking and non-smoking areas within buildings, although smokers and non-smokers did not differ in their symptom reports. Relative humidity was low in the buildings with a complete smoking ban, which may have prevented effects of either low RH or of ETS from being detected.

Hawkins & Wang (1991) conducted a cross-sectional study of 15 buildings, 6 of them naturally ventilated, 1 mechanically ventilated and 8 air-conditioned with humidification. All but 2 of the air-conditioned buildings had no-smoking policies. The buildings were not in polluted areas or known to be 'sick'. The questionnaire included a mixture of SBS, asthma and 'ergonomic' symptoms. Smokers reported more symptoms than non-smokers but there was no effect of passive smoking.

In none of the comparisons of buildings was there data on symptoms before a smoking policy was introduced, and extraneous factors were not controlled for. It is therefore difficult to judge their results. In general, it is difficult to show a link between active or passive smoking and SBS, as distinct from the effect of working near to a smoker. The presence of an obvious source of pollution may make workers more likely to report their symptoms or to attribute them to the obvious source.

Although there are some contradictions in the evidence, the current view of the WHO (1982) and indeed the UK Government is that there should be more vigorous efforts to curtail smoking, especially in public places. The justification for this depends little on SBS: ETS is only one possible contributor to SBS and the evidence for its effects is unclear at present, but SBS is by no means the most serious health implication of ETS.

3.3.4 Simple inorganic gases

(a) THE POTENTIAL PROBLEMS

Carbon dioxide levels due to respiration alone can rise to several thousand parts per million in well sealed buildings. Although there are no reports of levels above the occupational exposure limit (5000 ppm) levels of up to 1800 ppm have been reported (Janssen & Hill 1982) and since, in some cases, building ventilation systems have been controlled by CO_2 monitoring with the monitors set to operate at levels of up to 2,500 ppm, it is possible that CO_2 levels have, in some circumstances, approached the exposure limit.

Many other simple inorganic oxides and other gases can have irritant or toxic effects, and can be present in buildings - primarily as a result of combustion in or near the building, and less frequently as a result of the decay of building products.

(b) EVIDENCE FOR EFFECTS ON SBS

High levels of CO_2 can contribute to headache and lethargy but it is very unlikely that CO_2 at the levels found in buildings is itself a major cause of SBS or other illness. CO_2 is often used as an indicator for the presence of other airborne pollutants as noted above. However, such a standard will only be valid if there are no significant sources of pollution other than the occupants.

Kjaergaard et al (1990b) found, in a laboratory study, that the sensitivity of the eye to irritation by carbon dioxide is related to age, those under 40 being more sensitive, and reporting of SBS symptoms in the workplace, but not sex or smoking status. Thus there is a possible link to SBS here, but it is possible that the association between CO_2 sensitivity and SBS is due to an association between CO_2 sensitivity and sensitivity to other chemicals.

There is also little published evidence to suggest that combustion products drawn in through the ventilation system are a common cause of SBS. In a building investigated by BRE (Thomson Laboratories 1990) there was evidence of combustion products including sulphur dioxide re-entering the building from the flue serving a coal-fired boiler. Capping the flue to reduce re-entry of flue gases slightly reduced symptoms but the 'before' and 'after' stages were separated by a year and there were several other changes in the building over this period.

Various studies of airborne pollution in homes (see COST 1989) have suggested that nitrogen dioxide from combustion products may cause illness amongst children and some outbreaks of illness amongst building occupants investigated by HSE have been attributed to products of combustion from faulty heating equipment. However, there is no evidence to link heating or combustion equipment with published cases of SBS and such buildings are in fact unlikely to have unflued heaters.

Only one investigation links ozone from a photocopier ('wet type') with symptoms (Taylor et al 1984) although Skov et al (1989) found symptoms to be associated with photocopying work. Another report, involving a large number of buildings, discounts ozone from photocopiers. HSE investigations have shown pollution levels in photocopier rooms to be generally low so they appear unlikely to be a major cause of symptoms other, perhaps, than amongst some staff working in poorly ventilated photocopier rooms (Sykes 1989).

There have been specific reports from Sweden of SBS being caused in certain residential buildings by a mixture of ammonia and organic compounds from floor screeds. This work is described in greater detail in Section 3.3.4(b).

3.3.5 Volatile organic compounds (VOCs)

(a) THE POTENTIAL PROBLEMS

There is no doubt that a wide range of VOCs occur in buildings. Mølhave (1990) reports that there are 50 - 300 different compounds in a typical non-industrial environment. Most of them are well below the level at which they could have perceptible effects on people, but the mixture of 100 compounds, each 100 times lower than occupational threshold limiting values, could have acute effects which are similar in both type and time-course to SBS symptoms. Mølhave also describes the mechanisms by which VOCs can cause symptoms.

VOCs in the indoor air originate from numerous sources including furnishings, adhesives used in furniture and for sticking carpets, floor tiles etc, building materials, ventilation systems, building occupants and office equipment and materials. Thus, modern materials could partly account for the difference between air-conditioned and naturally ventilated buildings, since many of the former are of relatively recent origin.

This possibility is given greater weight by direct evidence of adverse reactions to building and furnishing materials. In a climate chamber study, Johnsen et al (1990) demonstrated break-up of the tear film of the eye resulting from exposure to rubber floor covering, acid-cured lacquered particleboard, water-based painted wallpaper on gypsum board and nylon carpet with rubber mat (a control group had no materials placed in the climatic chamber). This effect was attributed to the volatile organic compounds being emitted by the materials.

The VOC (strictly VVOC - very volatile organic compound) which is probably most frequently referred to in connection with indoor air quality problems is formaldehyde. This chemical originates from, for example, urea formaldehyde foamed insulation (UFFI), certain types of pressed wood board, carpets and tobacco smoking. UFFI has caused concern in domestic premises in some countries where measurements have revealed substantial levels (WHO 1982, Meyer 1984, Wanner & Kuhn 1984, Dement et al 1984, Matthews et al 1984) with peak values of 2.8 mg/m (Finnegan et al 1984) or more. Particleboard has been shown to cause higher levels when newly installed, dropping with time and ventilation rate (Meyer 1984, Sundin 1982). Formaldehyde is an irritant and may, therefore, cause symptoms similar to those of SBS.

VOCs may contribute to SBS symptoms at very low concentrations, and there are cases of individuals developing a sensitivity to formaldehyde (Johnson 1990). Formaldehyde and/or its metabolites are immunogenic: chronic exposure can cause activation of the immune system (Broughton et al 1990). But formaldehyde is not the only VOC of concern. The following are some recent examples of various sources in modern materials and a variety of compounds.

- An odour problem was found to be due to higher alcohols resulting from breakdown of phthalate plasticisers in carpet backing. The plasticisers were broken down by water, and calcium oxide in the carpet and alkali in the underlying concrete. Odour from the concrete persisted 9 months after removal of the carpet. McLaughlin & Aigner (1990).

- Some vinyl flooring materials contain, and emit at high rates for years, 2,2,4-trimethyl-1,3-pentanediol-di-iso-butyrate, a semi-volatile ester used as a plasticiser. Unconfirmed reports of reductions in SBS have followed removal of such flooring. Rosell (1990).

- The dominant source of VOCs from latex-backed carpets is the adhesive, rather than the carpet itself. Black et al (1991b).

- Coated parqueted floors are a long-term source of high VOC levels. Schriever & Marutzky (1990).

Gustafsson (1991) has reviewed more comprehensively the VOC sources among building materials and made recommendations for avoiding problems. There is similar activity in the USA (See 4.2.4).

But furnishings are not just primary sources of VOCs: they can also be 'sinks' - they can adsorb and re-emit pollutants. For example Kjaer & Nielsen (1991) demonstrated adsorption and desorption of VOCs on textile materials. Wool absorbs more than nylon and desorbs more slowly. Rates vary with compound, but desorption generally increases with temperature. Cleansers used on furnishings can also be a source of VOCs (Gebefuegi & Korte 1990).

Other than furnishings, there are several important sources of VOCs in buildings, as exemplified by the following.

- Tsuchiya & Stewart (1990) report that the VOCs found in different buildings make distinctive 'fingerprints' which indicate particular sources: these fingerprints suggested that wet process photocopying predominated in many of the buildings investigated.

- Hujanen et al (1991) assessed used filters in a laboratory study of odour level and odour acceptability. Dirty filters were rated less positively on both counts, particularly if they were taken from city centre buildings, and source strengths of up to 220 olfs were recorded. Unfortunately clean filters were not tested so it was not possible to say how much of the odour was the result of soiling. Nevertheless, the odour was almost certainly due primarily to VOCs.

- Virelizier et al (1990) note that personal cosmetic products can contribute to the load of VOCs. This is a rarely considered source, possibly because the chemicals involved are deliberately placed in the environment for a nominally positive purpose. There are, however, the beginnings of public concern about the air quality consequences of personal cosmetics, particularly from the point of view of persons who are sensitive to them.

- Gebefuegi & Korte (1990) found an air intake near the entrance to a car parking area to be a major source of VOCs.

- Exposure to VOCs can come from mains water (Tancrède & Yanagisawa 1990, Wilkes & Small 1990) which could enter the air via humidifiers.

VOCs can also originate from micro-organisms (see Section 3.3.5).

(b) EVIDENCE FOR EFFECTS ON SBS

Mølhave (1990) indicates a range of effects of total VOCs (TVOC) at different concentrations (see Table 5). The ranges for each level of effect are quite large; it is therefore not usually possible to make direct attribution of effects on building occupants from measured TVOC levels.

One limit on the potential for VOCs to cause SBS is the fact that levels would normally decrease over time following completion of a building. In one reported case (Ekberg 1991) levels decreased rapidly in the first 3 months of occupancy and were well within established guidelines and recommendations after a year. This fits with the report by Baldwin & Farant (1990) that levels decrease to a stable level over approximately a year. The level of VOCs at any one time would obviously depend on the starting levels and the nature of the sources, but the general principle seems to be that levels decay quite rapidly.

Table 5. TENTATIVE DOSE RESPONSE RELATION FOR DISCOMFORT RESULTING FROM EXPOSURE TO SOLVENT-LIKE VOLATILE ORGANIC COMPOUNDS (from Mølhave 1990).

Total Concentration (mg/m^3)	Irritation and Discomfort	Exposure Range
<0.20	no irritation or discomfort	the comfort range
0.20-3.0	irritation and discomfort possible if other exposures interact	the multifactorial exposure range
3.0-25	exposure effect and probable headache if other exposures interact	the discomfort range
>25	neurotoxic effects other than headache may occur	the toxic range

Thus, VOCs from building materials and initial furnishing materials become a much less likely cause of SBS once a building is approximately a year old. This is hard to reconcile with the finding that 1970s buildings are more likely to be 'sick' than 1980s buildings (Wilson & Hedge 1987). Nevertheless, VOCs are always worth considering as a potential case in new buildings, and there may be different sources and/or different causes in older buildings.

It has been observed that people adapt quite rapidly to the odour of air pollutants, and some have suggested that this means that odorous pollutants can be permitted at higher levels than previously thought (Gunnarsen 1990). However, this is not the whole story: the fact that people can adapt to odour does not mean that the adverse effects of a chemical disappear with the odour. For example, Hudnell et al (1990) found that the odour resulting from exposure to mixtures of volatile organic compounds decreased over time, whereas irritation of the eyes and throat did not. Thus the fact that people adapt to odours does not preclude the possibility that odorous substances contribute to SBS.

Against the above caveats, a number of investigations appear to have shown positive evidence that VOCs contribute to SBS. Among the earlier studies, Dement (1984) found complaints of upper respiratory and eye irritation in a temporary building with particle board panelling where formaldehyde levels of up to 0.23 ppm were measured. After 'fumigation' with ammonia and increased ventilation, formaldehyde levels dropped to below 0.1 ppm and complaints subsided.

Wallace et al (1991) found that persons in areas where new carpet had been installed during the year-long period covered by the survey reported increased frequency of throat symptoms (sore throat, dry throat, hoarseness) and possibly dizziness, but there was no association with measured VOC levels. Odour of paint and chemicals was associated with headache; nasal, chest and throat symptoms; fatigue; difficulty concentrating; and dizziness. Odour of cosmetics and odour of food were associated with eye symptoms, chills and fever, pains and difficulty concentrating. Some of these correlations are difficult to explain unless they are caused by statistical artefacts but this is possibly the only study which has considered cosmetics as a potential cause of SBS.

The same study found symptoms (headache; nasal, chest, eye, throat symptoms; fatigue; pain; difficulty concentrating; dizziness) to be correlated with self-reported chemical sensitivity. Of course, the respondents may have judged themselves to be chemically sensitive for the very reason that they were experiencing symptoms.

Berglund et al (1990a) studied the variation in 16 symptoms experienced by staff in a library building. Eight of the 16 were found to increase during the course of the day and these were reckoned to be the primary SBS symptoms. They do indeed include those symptoms which are reckoned to define SBS, except nose irritation and dry skin. The increase in these symptoms in the course of the day was very highly correlated with levels of total volatile organic chemicals.

Norbäck & Torgen (1987, 1989) studied the occurrence of SBS symptoms in the staff of 6 primary schools. In an initial cross-sectional study they observed an excess of symptoms in the two schools with wall-to-wall carpets compared to the four schools with hard-cover floors. The wall-to-wall carpets were eventually removed from the schools. A repeated questionnaire was directed four years later to those subjects still working in the schools. There was a significant decrease in symptoms among those whose carpets had been removed while the mean score in the reference group (those who had no carpets initially) did not change over time. The study was obviously not blind, and the conclusions would have been stronger if changes had been analysed within individuals, rather than by group comparisons.

Hellström et al (1990) eliminated spontaneous complaints in a school by removing PVC flooring and enlarging a flue but it is not possible to say whether the change was due to one intervention or the other, both, or the occupants' knowledge of the change.

In one major study (Skov et al 1897) there were two important correlates of SBS: the "fleece factor" (area of carpet, curtains and other fabric divided by the volume of the space), and "shelf factor" (length of open shelving or filing space divided by the volume of the space). These factors reflect possible sources of pollution such as organics, dust and microbiological elements. The researchers proposed that when the temperature is high and/or the relative humidity is low, or there is high work activity, the potentially allergenic material is released. This model is considered further in Section 3.3.6.

Several papers have attributed symptoms to a 'photochemical smog' caused by the action of ultraviolet light on organic contaminants but these are based on a single study of two buildings (Sterling & Sterling 1983a) and the theory does not appear to have been conclusively proven. It is considered further under "LIGHTING".

A number of studies have identified self-levelling concrete screed as a cause of SBS in some buildings, but these cases have all been in Sweden and mostly in apartment buildings. They are worth noting, however, as a warning against adopting particular building techniques in the UK.

Andersson et al (1990) report that a problem in a series of apartment buildings was due to casein in a screed partially decomposing to produce ammonia and 2-ethylhexanol (the latter probably from carpets). An intervention to remove the screed and replace it in a sub-sample of the apartments was successful in reducing the symptoms to the level established in apartment blocks without self-levelling screed. Over the same period there was no change in apartments which did not have the intervention. The study could clearly not be carried out double-blind, but the control group residents did believe (as a result of media information) that they had the contamination in their flats and this did not cause them to report high symptom levels.

Unfortunately, Bornehag (1991) found that the release of ammonia from self-levelling screed cannot necessarily be dealt with by removing the screed because ammonia may have diffused into the concrete below.

Stridh et al (1988) had earlier found that IAQ complaints and SBS symptoms in an office building were related to high recirculation air and self-levelling screed which had become damp. Varying the recirculation rate caused a change in spontaneous complaints and symptoms were correlated with the dampness of the floor on each storey.

Hellström et al (1990) eliminated spontaneous complaints in dwellings by covering a self-levelling screed with 2mm polyethylene sheet (and reducing ventilation fan noise) in dwellings.

Evidence for a role of VOCs in SBS has also come from the laboratory. Kjaergaard et al (1990a) exposed subjects to a mixture of 22 volatile organic compounds. 21 subjects were deemed to be healthy, while 14 had been reporting symptoms of SBS. The exposure level was 25 mg/m, a level far higher than would normally be found in office buildings (Finnegan et al 1984). The SBS group had higher changes in subjective ratings (of odour, air quality, irritation of the mucous membranes) increases in polymorphonuclear leukocytes in the eyes and changes in performance (as measured by a digit symbol test) when exposed to this mixture. The presence of polymorphonuclear leukocytes is an indicator of inflammation in an area of tissue.

In spite of the evidence for a role of VOCs, there are problems with an explanation of SBS in terms of VOCs, or indoor pollutants in general. Several studies (e.g. Hedge et al 1986, Robertson et al 1985, Skov et al 1987, Sterling et al 1983b, 1985), have examined variation in symptoms between buildings in relation to indoor air pollutants and the general conclusion has been that there are no characteristic differences in type of substances present or pollutant concentrations between 'sick' and 'control' buildings. De Bortoli et al (1990) found no association between volatile organic compounds and air quality complaints in office buildings of the European Parliament. However, SBS could result from the additive or synergistic effect of many pollutants (or specific patterns of pollutants), each of which individually is sub-threshold.

One problem of assessing VOCs as a cause of SBS is the wide range of compounds and mixtures of compounds found in the indoor air. One approach is to measure TVOC, although there are problems of definition here. Summary indicators such as the fleece and shelf factors may therefore be more useful measures in some cases than measurements of specific pollutants. Another approach which is currently under development is subjective evaluation of odour. This research has indicated that ventilation must be geared to removing not just the odour which is due to the people in the building, but also the (often greater) odour from building materials and ventilation systems (Fanger et al 1988). Alternatively, assessment of VOC levels may be assisted by categorisation as background, due to users, and outdoor (Berglund et al 1990b) or by rank order assessment (Morey & MacPhaul 1990).

3.3.6 Micro-organisms/biological pollutants

(a) THE POTENTIAL PROBLEMS

Biological airborne contaminants include pollens, spores, moulds, bacteria, viruses and miscellaneous by-products. Most pollens, spores and moulds are of outdoor origin but may accumulate or multiply within the office environment. Where there is no significant indoor source and the air filtration system is effective then the levels of fungi will be higher outside than they will be inside.

Robertson (1988) reported that of 223 buildings investigated, 34% contained potentially pathogenic or allergenic fungi, and allergic reactions to these organisms were demonstrated in the staff of a subsample of the buildings. He also gives many examples of malfunctioning or ill-maintained HVAC systems, with a variety of contaminating objects and dead animals. Extreme examples do not prove a thesis, but they certainly create a climate of concern which needs to be addressed. Apart from the organisms themselves, mycotoxins are found in the indoor air and can cause a range of symptoms including those of SBS (Schiefer 1990).

Any assertion that micro-organisms could have a role in SBS has tended to be linked with contaminated heating, air conditioning and ventilation (HVAC) systems, and particularly humidifiers in these systems. While humidifiers provide an obvious breeding site, there has in recent years been a greater knowledge of this risk and consequently greater probability of good maintenance although problems still arise.

An area which has more recently been the focus of research is air filters and there is growing evidence that they can be a breeding site for a variety of micro-organisms. Pasanen et al (1990) and Martikainen et al (1990) isolated a variety of bacteria and fungi from used filters, and only the bacteria levels were dependent on the water content of the filter. CO_2 was produced by the filters, at a maximum at 20°C and 95% RH and varying primarily with humidity. Fungal spores captured in filters germinated if there was sufficient moisture. The necessary moisture conditions exist in HVAC systems and typical metabolites (light aldehydes and organic acids) could be detected in used filters.

Another source of micro-organisms in HVAC systems is the ductwork. Porous insulation in HVAC systems can become a source of bacteria and fungi if it becomes wet and dirty (Morey & Williams 1990, 1991). Vacuum cleaning may be an effective remedial measure in some cases. Even without porous material, ducts are a potential source of fungal spores, especially if dirty and not sufficiently well insulated to prevent condensation (Jantunen et al 1990).

In a more detailed study, Valbjørn et al (1990b) showed that dust can collect in HVAC ducts, particularly at constrictions such as around heating coils or dampers. They therefore investigated 13 ventilation systems which had no cooling or humidification (they were therefore not able to examine interactions between dust and control of temperature/humidity). The systems were 5 - 29 years old, in schools and offices. Samples were taken from horizontal ductwork. The total deposited dust increased with age of duct (0.7 g per m^2 per year). The macromolecular organic part of the dust was between 0.9 and 8.9 mg per g of dust, similar to the level found in floor dust (Graveson et al 1990). There was no visible growth of microfungi, but 70 - 6200 viable organisms per g of dust (less than in floor dust). VOCs and odour production were both very low.

But the issue is not so clear cut. First, Sverdrup & Nyman (1990) concluded from a survey of 12 Swedish buildings that micro-organisms can breed in HVAC systems and be transferred to the air of the building, but that people are the main source of micro-organisms and the exhaust air contains higher concentrations than the supply air. Second, there are other possible sources of micro-organisms in buildings. Hansson (1988) notes that problems can arise in the structure of homes if the time allowed for certain structures to dry is too short. The same may apply in office buildings where wet construction (e.g. poured concrete) is used or if there is wet weather during construction: where there is moisture there is usually the potential for microbial growth. Micro-organisms can also grow on a wide variety of building materials, including particleboard, mineral wool and plasticisers (Ström et al 1990). Although these organisms are largely trapped in the material and do not become airborne, some are known to release volatile organic compounds which could have effects on the health of occupants.

(b) EVIDENCE FOR EFFECTS ON SBS

As suggested above, attribution of SBS to micro-organisms is not uncommon but the evidence presented is somewhat incomplete. For example Mainville et al (1988) specifically attributed symptoms of extreme fatigue in a Canadian hospital to mycotoxins from mould growing where water penetration had occurred. The range of symptoms was similar to SBS.

Downing & Bayer (1991) report that over 80% of buildings investigated (total of 35) were improved by changing operation and maintenance procedures, but they give no indication of how improvement was assessed or what environmental improvements were achieved.

As discussed in Section 3.2.3, there is some evidence that the use of a humidifier may be related to SBS. One study (Finnegan et al 1984) found SBS in buildings without humidifiers, but it was restricted to eye, chest and skin symptoms; another (Wilson & Hedge 1987) found the use of a humidifier (other than steam humidifiers, which are considered to be sterile) is associated with increased symptoms, although this effect was confounded by other factors.

These effects may be due to micro-organisms growing in the humidifier. Alternatively, biocides in the humidifier may have adverse effects or building humidification may permit growth of micro-organisms in other locations (e.g. in furnishings). It is not recommended that biocides be used in humidifiers.

Filters in air-conditioning systems can reach high humidity and support the growth of fungi (Elixmann et al 1990). Allergenic extracts from fungi in a newly-constructed hospital were used to challenge 150 patients said to be suffering from SBS. Allergic symptoms were caused in 135 cases. Although the patients were said to be suffering from SBS, this paper does not clearly define SBS in terms of the symptoms that were reported.

Nevalainen et al (1990), comparing building with a history of indoor air quality complaints and those without any complaints showed higher levels of airborne fungal spores, particularly mesophilic actinomycetes in the problem buildings. However, the nature of the complaints was not clearly defined.

Harrison et al (1990) studied 3 types of office building - naturally ventilated buildings, air-conditioned buildings with humidification and air-conditioned buildings without humidification. The latter group contained a high percentage of systems working via induction units. Relative humidity and temperature did not differ significantly among the three groups of buildings. Levels of particulates, fungi and bacteria were low in both types of air-conditioned building but higher in the naturally-ventilated buildings, although all levels were low by recognised standards.

Symptoms were highest in non-humidified buildings, lowest in the naturally-ventilated buildings and intermediate in the humidified buildings. Correcting for ventilation type, symptoms were correlated positively with bacteria and fungi but negatively with particulates. It should be noted that only viable organisms were sampled and it could be that filters and biocides increase levels of non-viable organisms or select certain species. Thus although airborne micro-organisms may contribute to SBS they appear to be a small factor compared with ventilation category.

Investigating the effect of changing the ventilation rate in 4 buildings, Tamblyn et al (1991) showed that reduced ventilation increased fungal spore levels in 3 out of 4 buildings. The fourth building had a less efficient filtering system and may therefore have been permitting the entry of high spore levels before the ventilation rate was increased. Weekly symptom rates were not significantly correlated with fungal spore levels. In two buildings in which control for building population was possible, the correlation was significant in one building but not the other. This discrepancy was not explained by the species present. On a case-control comparison of individuals, controls were exposed to slightly higher spore levels. All fungal spore levels were lower than those associated with clinical symptoms but comparable to previous reports of SBS (Harrison et al 1990; Holt 1990; Ström et al 1990). The negative results in this study may mean that one week is too short a period for such studies.

Others have also suggested that organic dusts from carpets can harbour organisms such as house dust mites that cause asthmatic attacks or that dampness can lead to microbial contamination which in turn causes illness amongst building occupants (Anderson & Korsgaard 1984, Reisenberg & Arehard-Treichel 1986). Danish studies have specifically suggested a link between the standard of office cleaning, the presence of organic dust in the office and SBS (Graveson et al 1990).

British research (Leinster et al 1990, Raw et al 1991) has sought to identify the causal effects responsible for this correlation by investigating the effect of improved office cleaning in a large office building.

In the first study, a cool shampoo technique was used to clean carpets, chairs and, where possible, other fabric surfaces. All hard surfaces (desks, tables, window sills, cupboards, filing cabinets and Venetian blinds) were then cleaned using wet techniques to remove dust. Files were vacuum cleaned in order to remove paper and other dust as thoroughly as possible. The carpet and chairs were also vacuumed following the cool shampoo. The vacuum cleaners used were all fitted with high efficiency final filters. The cleaning regime in the second experiment was similar, but with steam cleaning replacing the cool shampooing. The vacuuming of the files was not included but the rest of the activities were the same. In a control condition no deliberate changes were made to the offices.

Respondents reported a positive improvement in environmental conditions and SBS symptoms following the two cleaning treatments relative to the control condition. The overall reduction in weekly reported symptoms was greater in the second experiment (40%). Translated into annual symptoms for the building as a whole, this would mean a reduction from 4.2 to 2.5 symptoms per person. The final level is below the national mean for air-conditioned buildings (Wilson and Hedge 1897). This can only be an estimate of the magnitude of the benefits, arly since the sample size was not large.

: of interest is that the cleaning regime in the second experiment had a ...le effect on the presence of mites, fewer being found in rooms after steam cleaning than in the control rooms. This change was particularly marked for hard surfaces and seat covers. This does not prove that mites were responsible for the symptoms, but does raise the possibility. In this context, a comparison with the results of the first experiment is of particular interest since the cleaning regime used there was not expected to have a significant effect on the mite population: the overall reduction in symptoms was only 21%. There was also, incidentally, no effect of informing the occupants that improvements were to be carried out.

The implication of dust mites in SBS raises issues in relation to building services. The issues are raised here as questions, but discussion of them is beyond the scope of this report:

- does the relatively controlled and constant environment in air-conditioned offices promote the indoor viability of mites and micro-organisms generally;
- does the general standard of office cleaning need to be improved, possibly incorporating occasional steam cleaning and possibly using central vacuum cleaning to remove dirt from offices;
- does indoor air need to be filtered more effectively to remove airborne dust?

It has sometimes been supposed that mites would not be a potential cause of SBS in Scandinavia, where the low indoor winter humidities and occasional deliberate low indoor temperatures are expected to kill dust mites. Sundell et al (1990b) did however show that house dust mites can be found in Swedish homes, thus demonstrating that they can survive the Swedish winter.

In seeking to address problems caused by micro-organisms it should be remembered that the WHO (1990) concluded that the use of biocides in the cleaning and maintenance of HVAC systems or surfaces in buildings presents adverse health risks if the biocides become airborne in the building (see also Burge 1989).

3.3.7 Non-viable particulates and fibres

(a) THE POTENTIAL PROBLEMS

There are many sources of particulates and aerosols in the indoor air, including plants, animals, mineral fibres, combustion, home and personal care aerosols, and radioactive sources (Owen et al 1990).

Filtration media do shed fibres, but at a rate which reduces over time and still with a net benefit compared with unfiltered air (Shumate & Wilhelm 1990).

Office cleaning can also be an important source of internally generated particulates and is a function of the frequency, timing and quality of the cleaning techniques and equipment used. If the office is vacuumed daily, the timing of this will have an effect on occupant exposure. Dust stirred up by vacuuming is more likely to settle out before the offices are occupied if cleaning is carried out in the evening whereas there might be a significant residual level of airborne dust during working hours if vacuuming is carried out in the morning.

The type of final filters fitted to vacuum cleaners can have a significant effect on the levels of particulates which are generated during cleaning activities. Smith et al (1990) found that a standard vacuum cleaner emitted on average 9.1×10^{10} particles per minute in the range 0.01 - 10 microns during vacuuming. A survey of schools (Abildgaard 1988) showed that there is a high correlation between airborne levels of bacteria and dust. The lowest levels were in schools without carpets, the highest levels in schools with old carpets and high levels of settled dust.

(b) EVIDENCE FOR EFFECTS ON SBS

Most early papers are speculative and some give more opinion than fact about the role of dust. A few cases of 'temporarily sick buildings' may have been caused by airborne dust (Brown Skeers 1984, University of Washington 1982), and some authors have attributed SBS to dust and fibres from carpets, furnishings and insulating materials (Reisenberg & Arehard-Treichel 1986). There is, however, some stronger more recent evidence.

Wallace et al (1991) conducted a cross-sectional study of 3 buildings in Washington DC. Principal components analysis of questionnaire data generated 12 symptom clusters, 4 comfort factors (hot, stuffy air; dry air; cold, draughty air; humid air) and 6 odour factors (paint and other chemicals; cosmetics; tobacco smoke; photocopying/printing processes; new carpet; musty/damp). 68 regression analyses were conducted, each of 20 - 120 variables, the researchers therefore used a more conservative criterion in statistical tests ($p<0.01$). This is not necessarily the best statistical approach since it is 13.6 times less conservative than using $p<0.05$ for a single test.

By far the most frequently significant variable was airborne dust. All the SBS symptoms, all four of the comfort conditions and all six of the odour factors were strongly associated with dust. At first glance this would appear to make dust an important factor. There must be some concern however that dust levels were correlated with perception of 'cold, draughty air'. This suggests that to some extent dust was important because it was correlated with other environmental variables. Even so, about 20% of the employees at one of the buildings also brought up problems of maintenance (including lack of dusting or vacuuming, presence of vermin, etc) as the first item they mentioned on a free response question.

In the Danish Town Hall study (Graveson et al 1990) macromolecular organic dust was found to be correlated with mucosal and general symptoms. Dust was taken from floor dust samples. This effect could be due to immunological effects of the biological material or chemical, toxic or mechanical effects or some combination of these. Four buildings from the Danish Town Hall study (Skov & Valbjørn 1990) were selected for a follow-up study. This was carried out one year after the original study. Indoor climate parameters were similar to those originally measured and generally within accepted limits. Symptoms were more prevalent where the workers reported that the standard of office cleaning was not good.

Hedge et al (1991b) describe a system of 'breathing zone filtration' (by filters incorporated into office furniture) which reduced particulates and SBS symptoms and improved perception of ventilation, IAQ and thermal comfort in an office building. Such a system provides an alternative to increasing ventilation; it consists of a three layer filter: polymer pre-filter, activated carbon and high efficiency particulate air filter.

Unfortunately the 'control' in this study was retention of the old furniture and the comparison is therefore not perfect. Also, conditions were normally within ASHRAE IAQ standards throughout and there were similar temperature and humidity levels in both groups, but temperature was reduced in 'after' condition so both groups reported some improvement.

Some authors have attempted to link SBS with insulating materials associated with building services and heating and ventilating systems, particularly with asbestos (Bishop et al 1985), but this is purely speculation and none have justified their views.

One problem with an explanation in terms of dust is that the presence of air filters on recirculation systems has been found not to affect SBS (Wilson & Hedge 1987) although it is not clear what size particles would have been removed by the filters and it was not determined whether the filters were clear. Additionally, in another study (Hedge et al 1986) the level of particulates did not differ between two buildings, one of which exhibited SBS. Furthermore, Harrison et al (1990) point out that while adverse health effects are more commonly reported in buildings with HVAC systems, levels of airborne particulates are generally lower than in naturally ventilated buildings providing the air conditioning system is properly maintained.

Against this, it should be noted that the level of dust to which office workers are exposed can be 4-5 times higher than ambient airborne levels (Raw et al 1991). This is because people create their own 'dust cloud' in the course of their work by stirring up settled dust. Dust transferred direct from surfaces to the hands (and from there to the face), or ingested with food or drink, would be additional potential problems.

3.4 HYGROTHERMAL FACTORS

3.4.1 Summary

The levels of temperature, humidity and air movement required for comfort are well understood relative to many other aspect of the indoor environment, and the appropriate levels are generally specified and incorporated into designs. In spite of this, many buildings provoke complaints about being too warm, too cold, too dry, too humid, too draughty or too stuffy (which incorporates temperature, humidity and air movement in addition to odour). The complex way in which these factors interact to produce sensations of thermal comfort and dryness may be responsible for some of these complaints, or the design may not allow for individual variations in people's preferences, or the building may simply not be built and operated according to design. Many of the complaints are about not having personal control: over temperature in particular, but also the other factors.

The complaints described above are correlated with the occurrence of SBS symptoms, but are they causally related or are they independent consequences of poor building designs? The case for draughts being a causal factor is weak, but relatively few studies have been done. Controlled trials have shown that humidification to raise the relative humidity to above approximately 30% can reduce symptoms, although this benefit must be balanced against the possibility of humidification promoting the growth of micro-organisms. In any particular building and location it is also necessary to give consideration to the likelihood of humidity dropping below 30% for any length of time. On this basis, low humidity is less likely to be a problem in the UK than in Finland where the

definitive studies have been conducted. Longitudinal studies in Finland have also shown that symptoms increase as the temperature rises above 21°C. Although this result was obtained at a low humidity level, it does demonstrate that temperature can be a major factor at least in some conditions.

Temperature, humidity and air movement must be considered as determinants of indoor air quality, not just of comfort, since they can alter the rate of emission and deposition of pollutants from materials in the building and from people. These aspects are considered under the heading of indoor air pollutants.

3.4.2 The potential problems

Hygrothermal factors include temperature, humidity and air movement, all of which affect the comfort of building occupants.

Various standards have been set for the optimum comfort, the most widely accepted being that derived from laboratory studies and used as the basis for ISO 7730-1984 (ISO 1984). As with most such standards, this sets an optimum range of temperatures (air, radiant, and radiant symmetry) for people at different metabolic rates and wearing different clothing. Although the work is based on sensory perception it draws up complex equations to allow for the calculation of the 'operative temperature'. Recommended comfort requirements, from ISO 7730-1984, are:

- operative temperature 20°C-24°C (22°C±2°C) - with heating only systems, one figure would be chosen, but with air conditioning systems two figures could be selected, the higher one for summer (cooling) conditions;
- vertical air temperature difference 1.1 m and 0.1 metres (head and ankle height) less than 3°C;
- floor surface temperature 19-26°C (29°C with floor heating systems);
- mean air velocity less than 0.15 ms^{-1};
- radiant temperature asymmetry (due to windows etc) less than 10°C;
- radiant temperature asymmetry from a warm ceiling less than 5°C.

The standard is based on the 'predicted mean vote' (PMV) and the 'predicted percentage dissatisfied' (PPD) - and predicts conditions which are most satisfactory to most people for most of the time. 'Ideal' conditions can vary from population to population (surveys have shown optimum temperatures ranging from 17°C in Britain in winter to 37°C in the Middle East in summer) and from person to person within a population. No single thermal environment is ideal for everyone; even if conditions are 'ideal', a percentage of occupants will be dissatisfied.

The CIBSE Guide (1991) recommends that humidity should be maintained between 40 and 70%. Below 40% RH, unpleasant electrostatic effects may be experienced in modern offices due to the nylon and other plastics present in fittings and furniture. Above about 65% RH evaporation from the skin is inhibited and a feeling of stickiness or sultriness may be experienced.

Given these standards, there are several reasons why complaints about hygrothermal conditions do recur in SBS studies:

- older air-conditioned buildings are rarely provided with local temperature control;

- modern buildings do sometimes have a much higher degree of local temperature control but the level is governed by the relative quality of individual buildings and the cost constraints: building services designers would always choose to have individual control wherever possible;
- before occupancy took place, the controls (and the services plant) may not have been fully checked, commissioned and set up to ensure that the system met the design intent;
- controls may be hidden/inaccessible;
- maintenance has not been carried out in accordance with the operating and maintenance manuals provided at the end of the construction phase of the project;
- there may be a design fault.

Dissatisfaction with the thermal environment is often a greater problem in large air-conditioned buildings than in smaller and naturally ventilated buildings. The standards set in ISO 7730-1984 are complex and not easy to achieve in their entirety. In a building with opening windows and radiators the occupants are able to vary the thermal environment to some extent, whereas if the air conditioning or heating system in a large, 'tight' building fails to control the thermal environment, there is often little that the occupants can do to improve conditions.

A sensation of 'stuffiness' may also play a part. Stuffiness generally indicates dissatisfaction with the environment. Bedford (1974) attributes stuffiness to lack of stimulation, suggesting that a change of air velocity will stimulate the nerve tactile endings in the skin, which fits in part with the theory of Berglund et al (1984) that imperceptibly small stimuli may be to blame. Two papers report investigations where it was possible to reduce complaints of stuffiness by the use of individual fans to increase the air velocity from 0.05 to 0.60 metres/second (Int Hout 1984) and by reducing air temperature by 2°C from 23°C to 21°C (Matthews 1985).

Martin (1991) describes how carpets can cause or solve indoor climate problems by reducing thermal storage; maintaining differences between air and surface temperatures (particularly in lightweight buildings); retaining moisture in cool conditions and releasing it rapidly in hot conditions; and causing static shocks. Thus, reports which implicate carpets in SBS (see 3.3.4 (b)) may point to factors other than VOCs and dust.

Air movement which is too great can cause draughts. Fanger et al (1987) found that the percentage of people dissatisfied due to draught was a function of both the mean velocity and the turbulence intensity of the air movement. Air flow with high turbulence causes more complaints than the same air flow with low turbulence. In order to avoid sensation of draughts, Mayer & Schwab (1988) recommend that the flow of air should have low velocity (<0.3 m/s), low turbulence (<5%) and be in the direction of ceiling to floor. The second best flow direction is horizontal from the front, but never onto the back of the neck. In reality, people will change their working position so flow should not be horizontal unless people can vary the direction. Even then, care needs to be taken to avoid dry eyes.

Although there have been these valuable advances in understanding perception of draught, a physical description of the climate is unlikely to provide a complete explanation of occupant responses. This is exemplified by the observation that what may be considered acceptable with open windows in a naturally ventilated building may be considered to be a problem in an air conditioned building.

Possibly draughts from an open window are expected, whereas those in air conditioned rooms are often said to be localised, i.e. round the legs or neck. Also occupants moving into air-conditioned premises may have high expectations about their new environment, but are rarely given an explanation of what an air-conditioned office is or how it works, how different it will feel and how much control there is over their local environment. Finally, designers have a fairly limited range of choices in terms of air velocities for comfort and are normally careful about their selection, which needs to be related to the final air distribution and discharge terminals. If the designer follows the rules, and complaints still occur, the most likely reasons are:

- the occupants were not expecting air movement;
- the distribution of air is incorrect due to incomplete commissioning;
- the building was occupied before the plant was fully commissioned, particularly in terms of air balancing and correct air discharge quantities;
- the control system is not adequate;
- after commissioning the plant has not been properly maintained;
- the complainant is part of the statistical sample falling outside the normal distribution for satisfaction.

It has been suggested (e.g. Cooper 1982) that the relatively invariant environmental conditions are not ideal conditions, in other words that the controlled environment which is the goal of air conditioning is not as valuable as has been presumed.

Relatively little is known about the effects of fluctuations in the thermal environment, but Wyon et al (1973) found that small rapid fluctuations decreased work rate and accuracy, caused sleepiness and fatigue, and resulted in perception of decreased air freshness. Large, slow fluctuations, in contrast, simply reduced thermal comfort. A constant thermal environment seems to be preferable to either type of fluctuation. Whatever the explanation for this result, the implication for SBS is unclear unless we can establish where affected buildings fall on the scale of fluctuation. They may be subject to small fluctuations in temperature and are this reason be perceived as uncomfortable.

3.4.3 Evidence for effects on SBS

Valbjørn & Kousgaard (1984) reported temperatures above 23°C to increase the occurrence of mucous membrane symptoms but not headache in a Danish office building study. In a Canadian cross-sectional office building study, periods of high temperature and low relative humidity with imperfect ventilation were suggested to be contributory factors to typical SBS symptoms, although the effects of the temperature were not clearly reported (McDonald et al 1986). The results of laboratory studies (Andersen et al 1973, Rasmussen 1971) as well as of two epidemiological office building studies (Andersson et al 1975, Franzen 1969) on the human ability to sense the relative humidity are contradictory, although it is generally accepted that the increase of temperature (with constant relative humidity) increases the sensation of dryness in the range of indoor conditions.

Symptoms are generally correlated with perception of hot, stuffy or dry air (e.g. Wallace et al 1991, Wilson & Hedge 1987) but sensation of dry air is not always correlated with relative humidity (e.g. Sundell et al 1991): it can also be caused by dust and gaseous air pollutants. However, others have failed to link symptoms with comfort and in some cases SBS has remained even when problems of discomfort have been dealt with.

In Germany and Denmark in particular there has in recent years been some interest in the health consequences of draughts. While it has been shown that substantial numbers of office workers suffer from draughts (Kröling 1987, 1988), particularly in air-conditioned offices, no definite link has been established with SBS.

The case for an effect of humidity is stronger. Although Robertson et al (1985) showed similar relative humidities in 'sick' and control buildings, low humidity is known to cause some of the symptoms noted in 'sick' buildings. Erythema may be caused by low humidity; Griffiths & Wilkinson (1985) quote an incidence of itchy erythema of the face and neck amongst factory workers which was cured by raising the ambient humidity from 30-35% to 45-50%. McIntyre (1975) was able to demonstrate some increase in eye irritation at very low humidities (20%), presumably due to drying effect.

Experimental studies in Finland have shown a correlation between SBS symptoms and humidity and temperature. Reinikainen et al (1988) compared two areas of an office building, with RH at 45 - 55% in the humidified areas compared with RH as low as 10 - 20% in the non-humidified areas. SBS symptoms were lower in the humidified areas, especially dry throat and dry nose. There was less effect on skin symptoms, no effect on eye dryness and no effect on humidifier fever symptoms or respiratory infections.

In a follow up study, Reinikainen et al (1990) carried out a cross-over study to investigate the effects of humidification. Under free-running conditions the relative humidity was 20-30%; with humidification using a steam humidifier it was 30-40%. Humidification reduced symptoms of dry skin and dry eyes while irritation of the throat was non-significantly decreased. There was a significant decrease in allergic-type symptoms which consisted of nasal congestion and sneezing. Sensation of dryness was reduced but stuffiness was increased. This latter result may be partly due to the half degree rise in temperature in the humidified periods. Indoor levels of bacteria, moulds, formaldehyde and dust were very low throughout the experiment and did not differ between the humidified and non-humidified periods.

This study, although well-controlled, was only a short-term trial. In the long term there are potential problems with raising humidity levels if doing so encourages the growth of micro-organisms. A study carried out in Britain (Smith & Webb 1991) provides important complementary data. Symptoms were monitored over the two winters, before and after the installation of steam humidifiers. Symptoms of lethargy, blocked/stuffy nose and dry eyes were reduced, although headaches increased. While the study was not sufficiently well-controlled to say that the changes were due solely to the installation of the humidifiers, the data do at least indicate that there need not be major problems with clean humidification systems, even though increased airborne micro-organisms were found.

In a study by Palonen & Seppänen (1990), indoor climate parameters were varied in five buildings in Helsinki while repeated questionnaire surveys were conducted. The effect of temperature was analysed in detail for one building. SBS symptoms correlated highly with temperature (Jaakkola et al 1989) even among those who were satisfied with room temperature. The range of temperatures was 20 - 26°C. Room air temperature was highly correlated with number of external walls in the room (more external walls meant a lower temperature). There were many complaints about draughts, but air velocity was typically 0.5 m/s. Complaints were largely explained by temperature (possibly a radiant effect due to windows) and gender.

In a summary of their various experimental and epidemiological studies, Jaakkola et al (1990b) show temperature to be a major factor in the building concerned, symptoms increasing by approximately 25% between 21 and 25°C, with humidity an important additional factor. They recommend that the room temperature be kept at 21±2°C and the relative humidity above 20%.

It should also be noted that the association with temperature was found in extremely low relative humidity (10-20%) but with adequate ventilation rates; the effect of temperature may be different under other conditions and in particular may be reduced when the humidity is within the range more normally achieved in the UK (30-50%). The effect of temperature would also depend on the potential for emission of VOCs, since this increases with temperature.

The results demonstrate an overall benefit from using a humidifier, even if it is only to increase the relative humidity from 30 to 40%. This is important because using a humidifier to raise the humidity further than this may have adverse environmental effects because of increased microbial growth, and on condensation problems in the building. This result must be contrasted with conclusions from UK studies that water-based humidification may slightly increase symptoms. This is not too difficult to understand when one considers that the humidity is usually not so low in British buildings. Benefits of humidification, particularly clean steam-humidification, are likely to be greater in Finland where the study was carried out.

However, Abbritti et al (1990) report an early case of SBS in Italy where the researchers judged that lighting which was too strong, low humidity and the presence of fibreglass in the air were responsible for the symptoms, although a range of other characteristics were not assessed. Hence problems with low humidities may not be limited to the more northerly latitudes.

Although SBS is strongly associated with complaints of physical discomfort and lack of control over the indoor environment, it is doubtful that SBS could result solely from either thermal discomfort or lack of control. It seems likely then that if there is any relation between SBS and comfort complaints this is due partly to ill-health exacerbating discomfort. Furthermore, two important studies (Hedge et al 1986, Robertson et al 1985) found no differences in air temperature, radiant temperature, or air velocity between 'sick' and 'control' buildings, although the average values reported may be misleading.

3.5 LIGHTING

3.5.1 Summary

The main aspects of lighting which have been considered as determinants of SBS are the provision of daylight and the quality of artificial lighting provided in offices. There are increasingly sophisticated calculation and modelling methods available to aid design for either natural or artificial light, but many existing buildings have been designed with a relatively simple approach based on illuminance levels.

The presence of tinted windows is associated with a higher incidence of symptoms but this may be because of a correlation between tinted windows and a particular generation of buildings. Although natural light is usually said to be preferred, and people seated near a window tend to have fewer symptoms, the reasons for this have not been established. There is a little evidence that glare increases eye and general symptoms, and subliminal flicker from fluorescent tubes has been shown to contribute to headaches. The hypothesis that ultra-violet light from fluorescent tubes can result in the formation of photochemical smog has not been proven. There is therefore evidence that lighting can contribute to the eye and general symptoms of SBS but it is unlikely to explain the full spectrum of symptoms.

3.5.2 The potential problems

Lighting has the potential to affect health and comfort via average intensity, glare, flicker and spectrum. Visual environments which fall outside the generally accepted design recommendations for parameters such as illuminance or glare, e.g. those given in the CIBSE (1984) Code for Interior Lighting, are those most likely to lead to unsatisfactory conditions. Poor lighting and glare are known to contribute towards eye strain and headache. Surveys of indoor working environments have often elicited complaints about the lighting, with a general dislike of fluorescent lighting and a preference expressed for daylighting (e.g. Markus 1967).

Illuminance is specified in lux at the working plane and varies depending on the task being carried out. In normal commercial offices a standard service illuminance of 500 lux is common, with tolerance limits either side of the specified level within the area considered. Care needs to be taken in the terminology as minimum illuminance or average illuminance are apt to appear with less than clear definitions and both would depend on the ageing of the light sources and the quality of maintenance.

The basic lighting calculations for office designs use the surface reflectances of walls, ceiling and floor and the illuminance is therefore a function of the colour and finish of the surfaces. The illuminance increases with higher reflectance. Lighting has to be designed for most office situations reasonably early in the overall process, in order to meet contractual and installation obligations. The values of reflectances have to be agreed with the architect at this early stage, but his final decision on finishes frequently takes place much nearer the completion of the project and the reflectances may differ markedly from those used in the lighting calculations, normally to the detriment of the illuminance. What is less clear is whether the colour of the finishes contributes directly to SBS.

Glare, veiling reflections and daylighting are rarely specified in a project brief or specification. Veiling reflections in a well-designed lighting environment are unlikely to arise from glare reflected from paper on the work top, but great care needs to be exercised in the use of visual display units (VDUs) to minimise reflections from the screen caused by extraneous light sources. Among the solutions are task lighting, uplighting and combinations of the two. The increasing use of VDUs and less than ideal lighting solutions in the past, or currently in some situations, may be a contributing factor to SBS, but VDUs are considered later under 'electromagnetic factors'.

Spaces which conform to recommendations can still give rise to complaints about lighting. These may arise as a result of the lack of a complete understanding of the effects of various aspects of the luminous environment on occupant comfort. Particular areas of concern are lighting quality, temporal modulations and the presence or absence of windows.

Much research has taken place over the years to try to develop an index of lighting quality but the only quality index in use concerns the avoidance of discomfort glare. Other aspects of quality are still not generally considered as part of most lighting designs although they have been shown to have a strong influence on occupants' ratings of satisfaction with the lighting in a space. Flickering lighting in a working environment is at best a source of distraction and irritation and for some people it can cause acute discomfort.

Well maintained modern lighting produces insufficient visible flicker to trouble the vast majority of people. However, even though it may appear steady, all discharge lighting operated directly from the standard 50Hz AC mains supply, including that provided by fluorescent tubes, produces some modulation in the light output at 100Hz, i.e. every half cycle of the supply voltage. This 100Hz modulation is above the 'critical fusion frequency' for the human visual system and the light is not therefore perceived as flickering or pulsating.

However light which is pulsating, even if it is perceived as steady light, does not necessarily have the same effect as steady light. Studies of the electrical activity of the brain by Berman et al (1991) have shown responses to stimulations at this frequency, indicating that although the light modulation is not consciously perceived, certain areas of the brain are being stimulated. Eysel & Burandt (1984) have demonstrated that the pulsating light from a fluorescent tube affects the firing of nerve cells in the visual pathways. These are likely to influence cells in the area associated with the control of eye movements.

The presence of a window to admit daylight and permit a view out of the office appears to be important in producing a comfortable space, possibly as a result of the variations in lighting due to changes in external daylight. These changes are one of the major elements of visual interest in an indoor space.

In a study of windowless offices, Ruys (1970) found that 87% of the occupants indicated that they preferred to have windows in their office and that 47.5% of them thought that the lack of windows affected them physically and/or their work. Among the reasons given for being affected were: lack of daylight, poor ventilation, desire to know weather conditions, desire to look in the distance for the view, feeling of being cooped up, isolated and claustrophobic, feeling depressed and tense. It was found that size, office colour, lighting level, and the distance to a window had no relationship to the dissatisfaction with the lack of windows. Hence it would seem that the window in its own right performs a unique role, distinct from the provision of light. That unique role may be the ability to be in contact with the external world.

It is well known that sensory deprivation induces a variety of undesirable symptoms; studies by Magoun (1958) showed that for the brain to be alert and active it was necessary for there to be a minimum amount of sensory input. Schultz (1965) defines the concept of sensoristasis as "a drive of cortical arousal which impels the organism in a waking state to strive to maintain an optimal level of sensory variation". This balance may be upset by conditions of sensory restriction or sensory overload, creating a basic human need for an adequate variety and intensity of stimulation from the environment. If the sensoristatic balance is disturbed, e.g. by sensory restriction, disturbances in perception and cognition are likely. This may occur in spaces where the visual environment is highly uniform, as in some of the spaces found least satisfactory by Wilson et al (1987).

The need for contact with the external environment may possibly be traced back to the nature of man's primitive environment. Corth (1983) suggests that man is adapted to the forest, which means that he is attuned biologically to a particular type of light. Such an environment also offers spaciousness, a direct contact with natural elements such as air, light, warmth and coolness, and the facility to escape.

These are factors which Stone (1989) suggests may induce a need for those living in an enclosed environment to have unrestrained actual or potential access to the external world. The window, especially the controllable window, is the one element in the internal environment that offers, at the least, potential access to the external world. Stone goes on to suggest that without the window man may have serious problems adapting to his internal environment. Unfortunately, on this basis man should also want tinted windows since in the daytime they provide a means of seeing others without being seen. Tinted windows are among the common characteristics of 'sick' buildings (Wilson & Hedge 1987).

3.5.3 Evidence for effects on SBS

Various studies have been carried out to examine building occupants' subjective assessments of their luminous environment. These have mainly been aimed at defining measures of lighting quality and ranges of lighting design parameters to assist designers in the production of future schemes, although a number of studies have been specifically carried out where problems have arisen and corrective measures are being sought.

One of the largest recent studies of this type was undertaken in the USA by Collins et al (1989). They collected photometric and occupant response data at 912 work stations in a post-occupancy evaluation study. In addition to the more traditional considerations of lighting quality and control, they explored the concept of visual health. This was derived from reports of eye irritation and trouble with focusing eyes, which were found to be closely correlated, the former being a symptom of SBS. Visual health was judged to be poorer in those offices that were seen as either too bright or too dim. Lower scores on the visual health index were associated with lower scores on a lighting quality index developed from occupants, reports of lighting satisfaction, amount of light for work, amount of light for reading and the extent to which lighting hindered job performance. The response to glare was also found to be an important component of lighting quality.

People in work stations with daylight were more satisfied with their lighting than those who did not have daylight. This may be because of some intrinsic attributes of daylight, e.g. its lack of rapid temporal modulation, or the variations in lighting produced by the changes in daylight. Alternatively, the presence of a window together with its other associated properties such as contact with the external environment may be the crucial factor.

There are few comprehensive studies which have investigated the symptoms associated with SBS together with aspects of the visual environment. Robertson and Burge (1986) considered glare in their investigations and rejected it but subsequent studies have found more positive evidence.

A study by Robertson et al (1989) of two adjacent buildings, one air-conditioned and one naturally ventilated, also showed a general dislike for fluorescent lighting. There was a greater incidence of work-related headache and significantly more workers expressed dissatisfaction with the lighting in the air-conditioned building than in the naturally ventilated building. The workers in this building also found the light to be less comfortable and measured glare indices were higher. Measured illuminances on the desks in the air-conditioned building were low; 50% were below the CIBSE Code recommendation of 500 lux. Those workers with symptoms tended to have darker work positions. They also found the lighting less comfortable and perceived more glare. In addition they felt that their control of their own lighting and the office lighting in general was poorer than for those without headache. The windows in the air-conditioned building were smaller than those in the naturally ventilated building and were also tinted; the amount of daylight was therefore considerably reduced.

Wilson & Hedge (1987) and Wilson et al (1987) found the greatest number of symptoms among occupants of air-conditioned, open-plan buildings. These were also poorly illuminated buildings; the lighting was assessed as being highly uniform, artificial lighting levels were low, decor was dull and glazing was tinted, which reduced the amount of daylight entering the room. The perception of adequate daylight appeared to correspond with the reporting of a window view. Visual privacy and satisfaction with lighting had some impact on experience of symptoms. The authors note that lighting might have proved to be a more important factor had the buildings in the sample displayed a wider diversity of lighting conditions.

Hedge (1991) has since found evidence that lensed-indirect lighting is preferred to parabolic lighting and results in fewer eye symptoms.

A cross-sectional study of 3 buildings by Wallace et al (1991) found glare to be one of the main environmental variables to be correlated with occurrence of symptoms (headache, eye symptoms, fatigue and difficulty concentrating).

Wilkins et al (1984) proposed a link between headaches, eyestrain and light modulation, based on neural inhibitory mechanisms. Wilkins et al (1989) studied the effect of fluorescent lighting on headaches and eye strain amongst a group of workers in an office building. Although the offices were not deep and had reasonable amounts of glazing, they looked on to narrow light-wells and received relatively little daylight. The weekly incidence of headaches and eyestrain reported by the occupants was compared under two illumination conditions: when the offices were lit by fluorescent tubes operated with conventional circuits (providing illumination that pulsated at 100Hz) and with electronic ballasts driving the tubes at about 32kHz and substantially reducing the 100Hz modulation.

The average incidence of headaches and eye strain was more than halved under the high frequency lighting. The mean incidence of headaches and eyestrain changed with the change-over in lighting, showing a reduced incidence under the high frequency lighting, although the number of people who experienced both lighting conditions was small. A few subjects suffered headaches or eyestrain frequently and they did so mainly under the conventional lighting rather than the high frequency lighting. Headaches tended to decrease with the height of the office above ground and thus with increasing natural light. This may in part be a reflection of the relatively steady nature of daylight, compared with the temporal modulations associated with the conventional fluorescent lighting, a factor which may also contribute to the frequently stated preference for daylight over fluorescent lighting.

Another aspect of lighting that has received attention is the effect of certain types of light on indoor chemical pollution. Several articles and papers (Ferahrian 1984, Dimmick & Akers 1969, Int Hout 1984, Hansen & Andersen 1986, Andersen et al 1985) have considered this point, postulating that ultra violet rays in light from certain types of fitting cause photosynthesis of chemical pollution to create a photochemical smog that causes the symptoms of SBS.

Sterling & Sterling (1983a) found that by simultaneously increasing the ventilation rate in the building and changing the luminaires to reduce ultraviolet light they were able to reduce the incidence of eye symptoms. Despite being unable to measure photochemical smog in the building concerned, they assumed that this was the cause of the problem rather than either the lack of fresh air or the type of lighting. The study would have been strengthened by a cross-over design, i.e. switching the experimental and control conditions. In any case the number of subjects in the experiment was small and the statistical analysis incomplete.

The buildings where the greatest number of the symptoms associated with SBS are found tend to have interior spaces which are large and have relatively little daylight. A number of studies of windowless buildings have found similar symptoms among the occupants. These symptoms are also akin to those appearing under conditions of perceptual deprivation. There is, however, some difficulty in separating the effects of lighting conditions from the effects of other building characteristics.

Although there is not strong evidence that lighting is a major cause of SBS, it could well be one of the lesser contributory factors. The symptoms associated with SBS are wide ranging and it is difficult to see how lighting could influence some, such as dry throat or runny nose, however others such as headache and eye irritations are more easily associated with the lighting conditions.

3.6 NOISE

3.6.1 Summary

Although the capacity of noise to cause annoyance, distraction, tiredness and headaches is accepted, the mechanism of determination is complex, depending on intensity, spectrum, temporal characteristics, perceived source and the necessity of the noise, all tempered by considerable inter-individual variation. Thus, the evidence that 'sick' buildings are generally no more or less noisy than 'healthy' ones must be considered to be open to more complex analysis. For example, noise from air conditioning systems (e.g. the hiss from induction units) may cause more problems that the intensity would lead us to predict because it is perceived to be an imposed and unnecessary noise, nothing to do with the work being carried out. This said, it is unlikely that noise is a major determinant of SBS except as a contributor to the total environmental load on the office worker. Similarly, infra-sound is unlikely to be a serious problem at the levels encountered in offices, but could contribute to general symptoms in some buildings.

3.6.2 The potential problems

Noise can cause headaches and fatigue and affect concentration, by a combination of its intensity, frequency spectrum, location, predictability and acceptability (i.e. is the source of the noise considered to be necessary, and who is responsible for it). For example even noise which has a relatively low A-weighted level can contain some pure tones which may cause irritation or other disturbances (Molina et al 1989). People also vary widely in their subjective response to a given noise situation. This is partly an inherent characteristic in that some people seem to be more sensitive than others, and partly a consequence of the fact that 'annoyance' can be caused by a number of factors and it is difficult to separate noise from the others.

However the noise environment is normally specified in terms of a sound (pressure) level in dB(A) for the internal environment, to be achieved against an external sound level (which will normally be much higher) and to ensure that internally created noise can be attenuated to achieve the specified environment. An upper limit of 45dB(A) would normally be applied in offices.

When considering interference with work, the effect of background noise on speech communication is an important factor. Normal speech is carried on at a level of about 60 dB(A), so background noise approaching or above this level will demand some raising of the voice. This is tiring and can lead to difficulty or mistakes, particularly when using telephone or radio communication. In general, a noise level below 60 dB(A) will not affect ability to concentrate, speed or accuracy of mental or manual work, although many people would find levels of close to 60 dB(A) annoying. Experiments tend not to show any significant acute effect on work efficiency for most people below about 85 dB(A), though almost anyone would find this very annoying and regard it as tiring (Sutton 1991).

In order to achieve specified indoor standards, the building fabric must be designed to attenuate the external noise so that the internal design figures are not exceeded, and the building services design and associated equipment selection have to be carried out to ensure that noise from the services also meets the specified conditions. The importance of the latter should not be underestimated since indirect transmission can be created by resonant effects from primary plant vibration, or from noise generated by water in pipework or, more commonly, by air in ductwork and at discharge orifices.

Infra-sound refers to sound waves in air with frequencies below the human audible range, which can be produced by air-conditioning plant in office buildings. The upper frequency limit is generally set at 20 Hz, and the lower at 1 Hz. There is a gradual transition from the audible zone to the infra-sound zone; it begins at 100 Hz, below which tonal sensitivity steadily decreases, and vanishes at about 20 Hz, although noise in the infra-sound region is still perceptible.

Low frequency noise, apart from its acoustical aspects, can have other physiological effects (see Ising & Schwarze 1982, Ising et al 1982 and Kröling 1983). Infrasound may cause dizziness and nausea, but this is reported to occur at levels above 120 dB. Physiological effects are thought not to occur below 100 dB but psychological effects can occur down to the level of perceptibility. Ising et al described low frequency noise in the region of the audibility threshold as a non-specific stress factor similar to audible sound. They found a statistically significant effect on responses to the statements "I have to try harder than usual in order to remember something", "I need some peace" and "I feel in need of a rest". There is a similarity between these problems and the general symptoms of SBS.

For all the potential problems of noise in offices, it is unlikely that noise would contribute to SBS other than by exacerbating the 'general symptoms', i.e. headache, lethargy, lack of concentration.

Another potential problem has to do not with the cause of SBS but with the consequences of certain proposed remedial measures. Removing possible sources of indoor pollution, such as carpets, soft furnishings and false ceilings could create a new set of problems because of the effects on absorption and reflection of sound. This must be considered when implementing remedial measures.

3.6.3 Evidence for effects on SBS

There is no direct evidence on the role of noise in SBS; many workplaces are at least as noisy as 'sick' buildings, without any reports of SBS symptoms. Noise may, however, contribute to overall stress levels among workers and thus exacerbate complaints.

There is also no direct evidence of high levels of infra-sound in 'sick' buildings, and surveys (see Field 1987) have failed to find any adverse physiological or psychological effects of infra-sound in buildings.

Infra-sound measurements were made in the buildings studied by Kröling (1987, 1988) in the frequency range 10 to 100 Hz. Levels increased by 10-15 dB during the operating periods of the heating plant, but did not exceed the audibility threshold in the pure infra-sound zone (below 20 Hz). However the audibility threshold in the air-conditioned buildings in the frequency range 30 Hz to 40 Hz was usually reached, and sometimes appreciably exceeded. Nevertheless, noise produced by the heating system was regarded as acoustically annoying only in extreme cases, because of the low sensitivity of the human ear to low frequency sound. Switching off a low frequency sound source, however, was consistently described as a relief.

3.7 ELECTROMAGNETIC FACTORS

3.7.1 Summary

Working at a VDU has been found to be correlated with experience of SBS symptoms but the reason for this correlation has not been established. One possibility is that VDU work is correlated with a more direct causal factor such as being restricted for long periods to a working environment which is generally poor. Alternatively, the screen may attract airborne dust to the vicinity of the user. It is highly unlikely that electromagnetic fields around VDUs are responsible.

The balance and concentration of airborne ions has not been clearly demonstrated to contribute to SBS although ion generators may have some benefit if they cause dust to deposit out of the air (there are other ways of removing dust).

Static shocks can cause skin symptoms if they are sufficiently frequent, but have not been shown to be correlated with SBS symptoms generally.

3.7.2 The potential problems

Most VDUs have a low equivalent electrostatic potential on the screen and a low intensity VLF magnetic field thought by some to have potential for biological effects (US Congress 1989). HSE (1983) considered these effects and discounts any link between radiation from VDUs and cataracts, miscarriages or facial dermatitis (erythema). A more recent BRE report (Naismith 1991) concurs with this view and also confirms that there is no clear evidence of adverse health effects from sources of electromagnetic fields in general (e.g. from lighting, photocopiers, electric heating systems). It remains possible that the charge on the VDU screen could create a more dense concentration of charged particulates around the face of VDU users and thereby contribute to SBS. Alternatively, either flicker (see 3.5.3) could play a role or the use of VDUs with inappropriate ergonomic provisions could contribute to general symptoms.

'Air ions' are molecules or atoms which have either a positive or negative charge. They are produced naturally by various means, or may be created artificially using an air ioniser. Small numbers of charged air ions exist in outside air, typically one in 10^{16} (American Industrial Hygiene Society 1987) molecules, but these become depleted indoors to perhaps one tenth of the outdoor level (Hawkins 1982). Several reasons are suggested for this depletion - loss of ions in metal ventilation ducts, water vapour, dust and smoke acting as condensation nuclei and the depleting effects of VDUs.

Air ions are claimed to have a number of effects on human physiology and health. For example, medical research papers from Israel claim that during 'Sharov' days, the increased positive ion concentration in air causes increased neuroticism which can be overcome in some patients either by the use of tranquillisers or by the generation of negatively charged small air ions (Assael et al 1984, Rim 1977).

Other researchers suggest that negative air ions act as catalysts to remove trace gases, kill micro-organisms, affect the dispersal of aerosols containing micro-organisms, and claim that they relieve anxiety in rats and mice and reduce the death rates in rats and mice injected with influenza virus (Krueger & Reed 1976). Negative ions have also been shown to affect the mammalian respiratory tract although the researchers doubted whether this would also apply to man (Krueger & Smith 1960, Hamburger 1960). Some papers have also shown that positive ions increase air uptake in exercise (David et al 1960).

3.7.3 Evidence for effects on SBS

There have been several reports of VDU workers experiencing skin symptoms (Lindén & Rolfsen 1981, Rycroft & Calnan 1984, Stenberg 1987, Wahlberg & Lindén 1988, Matsunaga et al 1988, Berg 1989) but no clear reason has been established for this. SBS and VDU/skin symptom case and referent groups did not differ in previous episodes of acne, atopic or seborrhoeic dermatitis, migraine or light-sensitive skin.

There is also evidence (Wilson & Hedge 1987) that those who work 6-7 hours per day at a VDU experience more symptoms than those who work fewer or more hours. The causation of this effect is not clear: it may be that those working 6-7 hours happen to be the workers who are not 'professional' BRE users, trained in ergonomic use of VDUs, but people who use VDUs for long hours as a consequence of their main work.

The possible association between SBS and VDU work may not be attributable only to VDUs - the workers concerned are probably among the least mobile and this may be part of the problem. However, if the effect were due primarily to mobility it would be expected that all the symptoms of SBS should be associated with VDU work whereas only certain symptoms appear to be so associated.

Considered selectively, the published information on airborne ions can be taken to show that negative ions have great benefits. Some suppliers of negative ion generators have shown great selectivity in their interpretation, claiming benefits for their products that include relief from bronchitis, hay fever, catarrh, asthma, rheumatism, headaches, colds, eczema, high blood pressure, palpitations, conjunctivitis and laryngitis, plus resistance to influenza, increased ability to concentrate and reduced fatigue (Ball 1982) and, of course, a reduction in the incidence of SBS (Hedge & Collis 1987). Some such claims have prompted action by authorities, particularly the US Food and Drug Administration (Ball 1982, USFDA 1980) who seized nine brands of generator for misleading claims between 1959 and 1967, and the Advertising Standards Authority.

There is certainly no clear evidence that low ion concentrations could be responsible for the range of symptoms found in SBS. While many claims are made for the benefits of ionisers, the evidence is not conclusive that there are health benefits to be gained from using ionisers.

The claim in the UK that negative ion generators have a beneficial effect on SBS is largely based on work by Hawkins (1982). However, in subsequent studies Hawkins failed to repeat his earlier findings (Hawkins & Morris 1984). Investigation by other researchers showed either that negative ions had no effect in 'sick' buildings (Krueger & Smith 1960) or that they had no measurable effect on human mood and performance (Hedge & Collis 1987). One investigation (Finnegan et al 1987a,b) gave higher negative ion concentrations in a 'sick' building than in the control building. Additionally there is no evidence that ion concentrations differ between buildings which exhibit SBS and those which do not (Robertson et al 1985).

Finnegan et al (1987a,b) also studied experimentally the effects of negative ion generators on SBS symptoms and perceived indoor air quality in a total of 26 subjects in five offices. The length of the study period was 12 weeks. During a reference period (4 weeks) ionisers were run without ionisation, then the ionisers were first activated in three of the rooms and after further two weeks in the other two rooms (remaining in each case on for the rest of the study period). A cross-over design was not used because of concerns that the benefits of negative ions might persist beyond the exposure period. The mean negative ion concentration during the exposure was over 13 times higher than during the reference period but no change in SBS symptoms was detected. The effect of two important extraneous factors, temperature and humidity, was taken into account.

There may be reasons, besides the obvious one, why tests on negative ion generators have failed to show a benefit. Chandraker & Benhama (1990) note that the performance of ionisers can drop off rapidly with distance from ioniser. Krueger & Reed (1976) suggested errors in observation caused by pollutants (including O_2 and NO_2 from the generator) and failure to earth the subject. Even if they were correct, their suggestions do not auger well for the practical benefits of negative ion generators.

It is difficult to prove beyond all doubt that ion generators have no benefit whatsoever, but the balance of evidence suggests that they are of little benefit for dealing with SBS. Ionisers are able to deposit dust out of the air, and this may explain why the findings on ionisers are variable. Their effect may depend on how much dust is in the air and what pollutants are adsorbed onto the dust. Dust levels would also tend to influence perception of the dryness of the air.

Experience of static electricity has been found to be unrelated to SBS symptoms (Wilson & Hedge 1987).

3.8 PSYCHOLOGICAL FACTORS AND INDIVIDUAL CHARACTERISTICS

3.8.1 Summary

SBS should not be attributed to hysteria and if it is attributed to stress then the reasons for the stress should be sought out and removed.

Although SBS symptoms are normally measured by self-completion questionnaire, this is for convenience and most of the symptoms can be demonstrated by more objective means and shown to be correlated with questionnaire responses.

SBS depends on a number of individual characteristics, being more likely to be reported by women; by staff employed in more routine, low-paid jobs and those with a history of allergy. These characteristics should not be seen as causes but as factors which make an individual either more sensitive to the environmental challenges which cause SBS, or more likely to be exposed to those challenges. Also, and probably at least as importantly, these individuals may be more likely to see the office environment as an important part of the job and more willing to report symptoms which occur.

3.8.2 The potential problems

A popular and often expressed view, especially among some of those responsible for 'sick' buildings, is that the causes are wholly or in part of a psychological nature: that the symptoms are psychosomatic.

Two commonly cited possible psychological causes of SBS are hysteria and stress. It is wholly inappropriate to attribute SBS to hysteria, which is a specific psychogenic illness with little in common with the symptoms of SBS. Stress is a psychological phenomenon which can cause or contribute to a range of symptoms from acute reactions to long-term effects via the immune and cardiovascular systems, but it should normally be seen as having an environmental cause. It is therefore more appropriate to deal with stress under 'Management and organisational factors'.

The mention of psychosomatic or psychogenic illness should not be taken to imply that somehow either the symptoms or the cause are not 'real'. The symptoms are real and some can be demonstrated by objective tests. Subjective and objective measures of dry eye symptoms are correlated. Franck & Skov (1990) gave two questionnaires to workers two weeks apart to assess eye symptoms. The responses were well correlated with each other and with an objective evaluation of tear-film breakup. Taking the mean of the two questionnaire responses resulted in an even higher correlation with the objective measure. Other symptom reports can be validated by a combination of observation and diaries (see Burge 1990). Thus the symptoms can be shown to be real, not imaginary.

Burge et al (1990b) validated a self-administered questionnaire by comparison with assessments in a medical interview which used the same questionnaire. The same people were involved in both assessments; there is therefore a chance of contamination of the results. However, there was concordance of over 75% for eye and throat symptoms, lethargy and headache. By contrast, only 31% of work related runny nose and 21% of flu-like symptoms were concordant. The medical interviewer identified an additionally 5% of work related symptoms that were missed by the self-administered questionnaire. All in all, the questionnaire seems to be comparable to the medical interview and may be superior, since more symptoms on average are reported on the questionnaire.

The questionnaire survey in this study followed an identical one 2 years previously. The building symptom index on the two occasions was highly correlated (r=0.92) and the six building concerned were in the same rank order, although in all buildings symptoms had decreased by a similar amount.

There must also, where there is an effect, be a cause, whether this is to be found in the office environment, the nature of the job or in the individual respondent. For example stress may be higher in air conditioned offices because of the open-plan office layouts often associated with this type of building, and the resulting lack of privacy, monotony or because of lack of control or claustrophobic effects resulting from sealed windows. Psychological factors of this kind are unlikely to produce the observed coherent pattern of symptoms.

Those factors which affect large sections of a workforce are not discussed here, although they would obviously have a psychological dimension to their effects. The discussion concentrates instead on the influence of individual characteristics. These characteristics should not be seen as causes per se, but as factors which affect susceptibility to the causes.

3.8.3 Evidence for effects on SBS

The importance of individual factors was emphasised by Hedge et al (1986). In a comparison of a naturally ventilated building with an air-conditioned building, the latter showed a higher symptom rate, as expected. Within the air-conditioned building there were, however, important differences among departments. In the absence of obvious environmental differences, a department composed mainly of female clerical workers reporting low job satisfaction showed higher symptom rates than a department of mainly male professionals reporting high job satisfaction. The latter department in fact showed a lower symptom rate than the naturally ventilated building. Both departments were served by the same air conditioning plant, but were on different floors.

This is consistent with the sex/job type effects shown in larger surveys, but although female clerical workers doing routine work are most likely to be affected, male managers can also be affected. In fact the individual characteristic which has probably most widely been shown to be associated with SBS is gender (e.g. Wilson & Hedge 1987, Skov et al 1987), and this appears to apply to all job categories. A higher rate of symptom reporting by females is not unique to SBS but is found generally in health surveys. Wilson and Hedge also found a clear association with job category, professional and managerial staff reporting fewer symptoms.

Skov & Valbjørn (1987a,b) found that symptoms varied with a number of individual and job characteristics, for example sex (women suffering more than men), job status and satisfaction, workload and photocopying or VDU work. After taking into account all these effects, there were still significant differences among buildings, thus supporting the concept of SBS. Environmental pollutants were within acceptable limits (with the exception of dust and electrostatic charge), and varied little among buildings.

Zweers et al (1990) grouped symptoms into 4 classes - skin symptoms, eye symptoms, nose/throat symptoms and neurological symptoms, such as headache and fatigue. The only factors to have a significant effect on all 4 groups were personal factors, i.e. gender, previous experience of allergy, job satisfaction and satisfaction with handling complaints. Skin symptoms were additionally significantly affected by not being able to adjust the temperature. Eye symptoms were higher among those who wore contact lenses or who worked at VDUs. Nose and throat symptoms were higher among the older respondents and those who could not adjust the temperature themselves. Neurological symptoms were higher where there were more than 10 people sharing the office, and among VDU workers.

The Northern Sweden Office Illness project (Stenberg et al 1990a,b, Sundell et al 1990a, Sandström et al 1990) is a major investigation of 6000 workers. The interpretation of the results must take into account the fact that building-relatedness of symptoms was 'established' by a direct leading question. Also, environmental monitoring followed the questionnaire study, therefore the symptom and environmental data did not refer to the same period.

The initial results show that, in comparison with men, women reported more symptoms, a poorer environment, less interesting work, less influence over working conditions but fewer cases of having too much work to do. There was a slight excess of symptoms among those aged under 40, having a personal history of asthma and those using a VDU but it is not clear whether these factors were analysed controlling for gender. Men had higher outdoor air supply but also higher fleece factor (see 3.3.4(b)), it is therefore difficult to account for the gender difference in terms of the indoor environment.

SBS and non-SBS individuals were defined and compared. They did not differ in total air supply rate, outdoor air supply rate, temperature or relative humidity, fleece factor or shelf factor (see 3.3.4(b)). VDU users with and without skin symptoms on the face also did not differ on these variables. Individual symptoms did differ but in no coherent pattern or direction, therefore there may not be a single syndrome (or a case control study of people across a range of buildings may confound all the effects when independent variables are considered one at a time).

Wilson & Hedge (1987) note the following factors which may explain the effect of job type:

- increased sensitivity to environmental conditions due to doing routine sedentary work;
- inability to escape for part of the day from the office;
- physical stress due to particular types of work (e.g. typing, close paper work);
- general quality of working conditions: space, enclosure, furniture;
- power to get conditions changed.

The symptoms of SBS have been reported in some of the studies described above to associate with presence of allergic disease or previous experience of allergy (see also Jaakkola 1986). Wallace et al (1991) found symptoms to be correlated with a personal history of allergy, together with workload and conflicting work demands. Workers with no college education also reported more symptoms but this is probably related to job grade.

Guidotti et al (1987) describes an outbreak of illness in a telephone exchange when 81 members of staff were taken ill. Air sampling results were negative and the outbreak was eventually attributed to an employee with a "military history of involvement in psychological operations". Smith et al (1978) describe three outbreaks of mass psychogenic illness in industrial plants with some symptoms similar to those of SBS. These outbreaks were found to affect workers in a predominantly female workforce who were under some physical and psychological job stress, and were triggered by a physical stimulus of some sort, most commonly a strange odour (Guidotti et al also mention a strange odour).

There are clearly some similarities between symptoms of SBS and those of psychological origin, but there is little evidence to link SBS with outbreaks of psychogenic illness of the type reported by Guidotti et al and Smith et al. Finnegan & Pickering (1986) conclude that symptoms are unlikely to be of psychological origin except in a small number of cases.

If SBS is not primarily psychological in origin, it has psychological consequences and may be magnified by psychological effects due to perceptions of the environment, the response of other building occupants and suggestion due to hearing about SBS. Psychological factors are generally more likely to affect the reporting of symptoms rather than the symptoms themselves. In short, the most simple model is that the environment causes symptoms and the reporting of those symptoms is dependent on psychological factors. Beyond this there are clearly individual differences in susceptibility and those may in time give clues as to the causes of SBS if the pattern of differences can be more clearly documented.

3.9 MANAGEMENT AND ORGANISATIONAL FACTORS

3.9.1 Summary

To claim that bad management causes SBS can be seen as either claiming an obvious truth (i.e. problems in the workplace are always due to bad management) or propagating an unfair nonsense (i.e. it is not the management, it is the environment in the building). The correct balance between these views can only be struck by establishing in specific terms what management could have done which would have avoided the problems. Broadly speaking, management can be seen as contributing to SBS if it does not act effectively to create a good indoor environment to avoid symptoms or if it does not establish a good organisational environment for avoiding stress and for dealing with complaints.

One clear correlate of SBS, which could be avoided by better management, is perception of lack of control over the physical environment or the work environment. Currently available evidence does not clearly establish whether this is because controllability is an important variable in its own right or if individual control gives the ability to individuals to create an environment which suits them.

3.9.2 The potential problems

The assertion that management factors contribute to SBS is based on two lines of argument. The first is that poor quality management can lead to inadequate environmental conditions which are a direct cause of symptoms. The second is that poor quality management affects the sensitivity of staff so that they suffer or report symptoms even in environmental conditions which would otherwise be considered adequate.

There are several ways in which management can fail to control those aspects of indoor environment which contribute to SBS and thus allow environmental problems to arise, for example:

- lacking data or simply not knowing enough about the environment and the building services, for example, what temperature and rate of ventilation are required;

- failing to maintain - for example, poor cleaning, not replacing air filters or luminaires which have failed;
- failing to set adequate standards - for example, low ventilation rates, not in line with CIBSE recommendations;
- failing to contemplate the consequences of change - for example, not changing the pattern of air distribution after space planning changes such as the introduction of perimeter offices;
- failing to intervene - for example, not commissioning equipment properly, not isolating or replacing noisy equipment.

People's sensitivity to environmental conditions is not static: their mental state (e.g. whether or not a person is dealing with a stressful work situation) and state of health can affect their physiological sensitivity. This means that people under stress can suffer more from deficiencies in the environment than those not under stress and so, since the quality of management clearly affects the general level of job stress experienced at work, it can be surmised that poor management will generally lead to greater sensitivity to the work conditions.

However, in addition to general job stress, there are other ways in which those responsible for buildings can cause additional stress for staff through management failure, for example if complaints of discomfort are ignored or not responded to sufficiently promptly. A BRE investigation (Leinster 1990) found evidence of a long history of staff complaints and a lack of co-ordinated management response and this is not unusual in such investigations. This may be why some buildings with high rates of SBS are reported by managers to have no problems, with the reverse also true (Wilson & Hedge 1987).

Rask & Lane (1989), in a secondary analysis of data on a number of buildings whose occupants suffered SBS, report that "psychosocial factors, such as employee distrust of management that resulted from a perceived no-action response to employee complaints" was one of the causes of SBS in these buildings, but the attribution of cause in this way must always be suspect.

If problems are not dealt with rapidly and effectively, staff can lose confidence in management and become frustrated and despondent about their environment. Besch & Besch (1989) state "the longer it takes to ... institute corrective action, the more likely the performance, productivity and morale of everyone ... will be unfavourably affected".

Research has also shown that staff doing routine, sedentary work, such as those at clerical and secretarial grades, are more prone than other staff to stress (Johansson & Aronsson 1984) and to increased incidence of SBS, as noted in the previous Section.

As Lindvall (1985a,b) has noted, "in offices, perceived discomfort is a critical effect" because of the mental nature of office work. Therefore, levels of environmental parameters which are well within the recommended limits can still be "psychologically stressing, increasing the risk of cumulative effects". It is commonly staff at clerical and secretarial grades who have the worst office environments, often without natural light and without the good environmental controls which are generally found in cellular offices. These staff have increased sensitivity because of the routine, sedentary nature of their work and it is therefore a task of management to address their special needs and provide a good quality working environment to compensate for their pattern of work.

One factor which increases the sensitivity of staff in many modern buildings is that they are provided with highly centralised environmental control systems which are often enhanced by design features which reject the external world through the use of tinted glazing and sealed facades. In such buildings, occupants can have little opportunity to exercise control over their environment and so they are totally reliant on the building systems and those managing them to produce a satisfactory environment (Kroner 1988, Wilson et al 1987).

Of course, where there is a perception of lack of control, this may be factually correct but there are many cases where various local controls exist and no instruction has been given to the occupants about their function or operation. This still amounts to lack of control from the occupant's perspective, but the solution to the problem will clearly be different.

The provision of local controls, particularly for temperature and lighting, is important and is now widely accepted. However, the brief and its required level of comfort, system selection and cost all affect the degree of local control. It is impossible at this time to claim either that the perception causes a grievance which contributes to SBS, or that the apparent uncontrolled parameters themselves are at fault. What can be identified are the possible options which cause this perception and, where relevant, their solutions.

One possibility is that the occupants are correct and there are no local environmental controls. This could be due to the age of the building, the quality of the environment specified for the building and/or design solutions constrained by cost limits. The remedial measures to provide such controls are normally expensive and sometimes virtually impossible to achieve. Alternatively the occupants' perceptions may be wrong, due to insufficient training and instruction as to how the building services systems operate, about what local controls there are and how they are set and function. The solution to this is relatively easy - suitable training needs to be provided by management.

The possibility remains that the occupants' perceptions are justified even though local controls do exist and they have been given suitable training and instructions as to their use. In this case the probable reasons can be selected from one or more of the following:

- the design may be at fault;
- the controls (and possibly the plant) were not properly commissioned;
- the controls are inaccessible;
- the commissioning was properly carried out, but routine maintenance and re-calibration of the controls have not been done.

Lack of personal control, whatever the reasons for it, tends to decrease people's tolerance of discomfort and to increase their environmental sensitivity and they may thus become more prone to SBS. It could be argued that the provision of de-centralised environmental control is unrelated to management quality; however, as outlined more fully in Chapter 4, management can influence the way in which environmental control is provided at the briefing stage of the building design process. It is also worth making the point that managers can recognise and compensate for circumstances in buildings where individual sensitivity is increased by lack of environmental control, for example, by being more responsive to staff complaints.

The notion of control extends to privacy (control over seeing and being seen, hearing and being heard) and being affected by the actions of others (e.g. closing windows, smoking, using chemicals). These aspects of control will obviously depend on the number of people sharing an office, but so does perception of control over heating, ventilation and lighting (Raw et al 1990). Good office management can also improve personal control in these areas.

3.9.3 Evidence for effects on SBS

There is a suggestion by some investigators that dissatisfaction with the working environment, especially with lack of control, and working conditions may play some part in SBS. Waller (1984) in a case study of a large air conditioned insurance office attributes complaints to frustration at lack of environmental control, poor management, discomfort and dislike of open plan offices. However, the paper does not mention medical symptoms. WHO (1982) also discusses complaints and negative attitudes towards air conditioning, again without mention of symptoms.

Wilson & Hedge (1987) consider the link between symptoms and employees' working environment, their perceived ability to control the environment and their job satisfaction. They found more symptoms in buildings where occupants had little perceived control over the environment (temperature, ventilation, lighting, noise etc) and the highest rates were in public sector 'clerical factory' environment. Perception of lack of control was associated with air-conditioned buildings, but unlike SBS, not with any particular type of air conditioning.

Management has the potential to affect SBS in direct and indirect ways, as noted above. It is difficult to prove an association but often relatively easy to suggest improvements.

4. WHAT CAN BE DONE?

4.1 INTRODUCTION

4.1.1 Summary

Until a more complete picture of the causes of SBS emerges, action should be taken on the basis of the knowledge we have. This will need to be directed in an integrated and multi-disciplinary manner to all stages in the life of the building, and to cover the building itself (and its location), the indoor environment, the organisations which occupy buildings and the needs of individual workers. A wide range of risk factors needs to identified and avoided or removed if SBS is to be eliminated.

4.1.2 An approach to finding solutions

Having examined the evidence on each suggested cause, SBS begins to look like an effect without a cause: no single factor has been identified as responsible for SBS and there are reasons to doubt each explanation taken in isolation. There is nevertheless an imperative to use what data we have to avoid producing buildings which will cause SBS, take what remedial measures we can in existing buildings and conduct new research which will give clearer guidance for the future.

It is helpful to view the causes of SBS at four levels:

- the building: many aspects of the design, construction and location of the building and its services and furnishings may contribute to SBS in a wide variety of ways from the site-dependent microclimate through shell design (i.e. depth of space, floor-to-ceiling heights, comfort conditions) to the services and fitting out;
- the indoor environment: the effects of the building and site must generally be mediated by the indoor environment;
- the organisations which occupy and operate the building: particular organisations may contribute to SBS, for example via the quality of building maintenance and work force management;
- the individual: reported experience of SBS varies from one person to another within buildings, for a number of reasons which would include personal control over the environment, constitutional factors, behavioural factors and current mental and physical health.

Interventions at one level may fail to resolve problems which arise primarily from failures at another level. Furthermore, the building is a system rather than a collection of isolated components; therefore, the effects of changes to one component are modified by the state of the other components. The co-ordination of each level and of architecture with structure and building services is therefore necessary to achieve a 'healthy' building which is both aesthetically and functionally suitable for the occupants' needs.

In the past the various strands of design have often been treated as separate elements for periods well into the design programme, but there is wide recognition now that modern buildings require a multi-disciplinary design approach starting very early on in the design process. As buildings and their services become more sophisticated this approach is essential. Clients, designers and building users who wish to reduce the incidence of SBS will:

- be more explicit in their environmental requirements in the project brief and design specification;
- design on a totally integrated basis: structure, materials, services and provision for operation and maintenance;
- closely monitor the building and installation to ensure that environmental co-ordination is achieved;
- ensure that operating and maintenance procedures are adequate to retain the design intent and operational performance over a sustained period.

This chapter is based on the principle that prevention is better than cure, and therefore starts with specification of the building and works through design, building and installation, commissioning and fit-out, to operation and maintenance of the occupied building and solving problems in existing buildings. The final Section focuses on the role of the building management staff. At each stage there is a requirement for communication and teamwork, management and planning, and quality assurance (Wenger 1991).

While it is difficult at present to be definitive about the causes of SBS, we can identify 'risk factors' and avoid them; they can be summarised as follows (those which have been more clearly shown to relate to SBS are in bold type).

The building:

- **open plan offices of more than about 10 work stations/deep building plan;**
- changes of use of the building and office partitioning after commissioning;
- sealed windows;
- **large areas of soft furnishing and open shelving and filing;**
- **new furniture, carpets and painted surfaces;**
- lightweight thermal properties and poor insulation;
- poor provision for daylighting and/or uncontrolled solar gain;
- no separately ventilated spaces for smoking, photocopying etc.

Services:

- **air conditioning (i.e. with cooling capacity);**
- humidification (steam probably a lesser risk);
- low fresh air supply/high recirculation rate;
- **low level of user control over ventilation, heating, lighting;**
- **not designed for easy maintenance;**
- poor air distribution within occupied spaces;
- air inlet close to exhaust or outdoor pollution source;
- inadequate filtration;
- **luminaire type and position giving high glare and flicker.**

Operation and maintenance:

- inadequate commissioning and, when necessary, recommissioning;
- **poor maintenance (hygiene and operation) of building services;**
- **insufficient office cleaning and suspect general repair.**

Indoor environment:

- **chemical pollutants, e.g. tobacco smoke, ozone, volatile organics;**
- micro-organisms from water, air and furnishings;
- **particles and fibres;**
- **high temperature or temperature changes during the day;**
- **very low or high humidity;**
- excessive or very low air movement;
- very low or high lighting levels, glare, **flicker;**
- high levels of noise (including infra-sound).

Job factors:

- **high levels of routine clerical work;**
- poor general management and management of staff complaints;
- low general satisfaction with the job and organisation;
- **aspects of VDU operation, which have yet to be specified in detail.**

Risk factors which are specific to the person are not included here because the focus of preventive and remedial measures should be the building, not the selection or medical treatment of individuals. Similarly, building age is a risk factor (1970s buildings presenting a greater risk than more recent buildings) but this is hardly amenable to modification. Neither is it known whether this is due to changes in design practice or to the degradation of buildings over time.

Discussion of these issues is expanded in the Sections which follow. There is existing guidance on most aspects (for example from CIBSE, RIBA, BRE); this report does not seek to duplicate this guidance but to highlight specific factors of importance which appear to have been overlooked in the past. In fact it is one of the frustrating characteristics of SBS that buildings which conform to current guidance and standards are apparently not immune to SBS.

Assessment of conformity with standards has not been a major element in SBS investigations, although failure to meet standards for air change rates is not itself a clear risk factor (see Section 3.2.3). Air pollution levels are typically well below UK occupational standards and physical parameters within normal comfort bands or no different from comparable 'healthy' buildings.

In short, the complex causation of SBS makes it difficult to specify standards parameter by parameter given the range of possible additive and synergistic effects. Clearly, if standards are not met in a particular building the situation should be rectified, but this in itself should not necessarily be expected to prevent SBS.

4.2 BEFORE BUILDING: SPECIFICATION, BRIEF AND DESIGN

4.2.1 Summary

Many professional groups, including building owners and managers, need to be involved in the construction and operation of an office building. If effective action is to be taken to avoid SBS in new buildings, all parties need to act in a co-ordinated manner, governed partly by current legislation and partly by guidelines and a desire to produce better buildings. The following risk factors in particular should be considered.

- Building design/location:
  - deep spaces, open plan offices;
  - poor daylighting;
  - local outdoor pollution.
- Building services:
  - the need for air conditioning
  - air supply (total volume, positioning of air intake, recirculation vs heat recovery, filtration, air distribution, hygienic components and the requirement for humidifiers);
  - air distribution;
  - occupant control over the systems.
- Selection of building and furnishing materials for low pollutant emission.

4.2.2 Who is responsible?

A desire to achieve integration of all facets of building raises the difficult technical, logistic and legal question of who is responsible for achieving the necessary integration. The important point is that someone should take responsibility, beginning with the specification and brief for the building.

For example, Vatne (1990) describes a system being implemented in Norway whereby owners of buildings are made more responsible for them. The main objective is to clarify who is responsible for the various elements of the final result:

- the owner has to specify what he wants to purchase;
- the consultant, in co-operation with the owner, has to ascertain that the correct acceptance criteria are included in the tender documents;
- the builder is responsible for compliance with the acceptance criteria;
- the owner presents the required documentation to the Labour Inspectorate, including the test results verifying the design figures.

Hence, the specifications will make it possible to determine who is responsible for unacceptable deviations, and a guarantee will have a real meaning. The items which have to be specified are listed in Appendix 1.

In the UK the HSE has suggested to the Parliamentary Select Committee (Environment) a more limited basis for which could in principle be applied to regulating the maintenance of ventilation systems, and could also apply to other building services. This comprised the following elements:

- that installers of forced ventilation systems be required to provide users with a statement of environmental parameters to be achieved in all parts of the building and a schedule of maintenance;

- that operators be required to operate the system to meet its design parameters, follow the schedule of maintenance and keep records;
- that occupiers also be required to have periodic examinations and tests of the system undertaken by a competent person.

This is unlikely to be adopted as a mandatory requirement, the reasons given by the HSE being cost-effectiveness, uncertainty about the nature of the role played by ventilation systems and the relative seriousness of SBS as a health issue for individual workers. This type of approach could, nevertheless, be recommended as 'best practice'.

Even without these requirements there is scope within the current Building Act (1984) to strengthen the role of the owners and occupiers, on whom the Building Regulations "may impose .... continuing requirements" (Section 2(i)). There can be little which is relevant to SBS which is not mentioned in the issues covered by the Act (Sections 7 and 8). In a more market-oriented approach, Prezant et al (1990) offer suggestions for contract clauses to promote good IAQ through occupancy agreements.

Some may wish for a greater influence of official guidance or legislation to be imposed on specifications. There is extensive guidance of certain aspects, provided by professional and governmental bodies, and this would normally be followed in good specification. The idea has been mooted of adopting more stringent indoor air quality standards in non-industrial workplaces. Although many countries have set standards for outdoor (ambient) air pollutants world-wide there are no non-industrial standards for indoor air pollutants (Seifert 1990). It is not the purpose of this review to resolve this debate, but it may be of value briefly to consider the contrasting views of two European experts.

Lindvall (1985a,b) points out the potential for occupational exposure limit values for offices to reduce air pollution in these spaces. Since office work for many is characterised by high mental performance demands, avoidance of toxicological effects is not an adequate criterion for a suitable environment. In offices, perceived discomfort is a critical effect: sensory warnings have a great emotional impact to many people and may cause exaggerated responses, even to buildings with only minor climate problems. They may cause unjustified claims of serious and persistent health effects. Increased sensory irritation, discomfort and fear of more serious persistent health effects easily add up to occupational health problems of a considerable size that makes preventive measures very cost-efficient.

Seifert (1990) on the other hand suggests three reasons against the setting of standards for chemicals and micro-organisms in indoor air:

- the existence of a standard favours the impression of having a limit below which there is no reason for concern;
- the enforcement of a standard is virtually impossible due to the large number of closed spaces;
- a large variety of conditions are encountered within buildings and a room may meet the requirements of a standard at 20°C but not at 23°C (both temperatures are within the comfort range and an individual cannot be constrained to live at 20°C simply because the air quality standard would be met only at that temperature).

He therefore prefers the use of guidelines. He considers that these are less binding than standards and are more likely to be agreed upon by different countries, in contrast to standards, because they generally will not take into account socio-economic or political aspects. A strategy that includes two values can be adopted. While the first value would define the level of hazard, the second would indicate a target value for the future.

This strategy can become quite complex: Seifert proposes a guideline figure for total VOCs of 0.3 mg per cubic metre. Compounds are assigned to one of seven groups and the following conditions must also be observed: no individual compound should exceed 50% of the concentration allotted to its own class or 10% of the total concentration. This proposal is not based on toxicological considerations but Seifert concludes that it is achievable taking into account current findings in the literature. The proposal, he also states, does not exclude the fact that additional guideline figures may be needed for individual compounds. The figures recommended are also intended for 'normal' operating conditions. If renovation work has been carried out, then for periods of 1 week and 6 weeks, 50 and 10 times these values would be acceptable.

In reality, all parties involved in producing a new building have a responsibility for their own part of the process; the issue which is to be resolved is who should be responsible for co-ordinating the whole effort and how will the co-ordination effort be resourced? It is the eventual occupier who has most to gain from a 'good' building, but does the occupier necessarily have the expertise available to perform the role? The evidence presented in this review about the causes of SBS does not answer these questions, but they must ultimately be resolved if better buildings are to be created.

4.2.3 The building form and location

Deep building spaces require more complex services than shallow space and so in turn requires more sophisticated management. Decisions about space are therefore of fundamental importance and must be made in relation to the resources available for services management and maintenance. Deep buildings also tend to employ open plan spaces, another risk factor.

Properly controlled daylight is usually the preferred form of lighting as it offers building users greater variability and stimulus than artificial lighting (Boyce 1981). Daylit spaces are more feasible in shallower plan buildings; however, light-wells and atria can bring light into deeper plan offices.

A factor which is often forgotten in SBS research is the building's location. Levin (1990) recommends avoidance of areas polluted by, for example, vehicle fumes, power generation, other combustion and industrial processes, agriculture, wind-generated soil erosion. Docks and transport depots can also be sources of pollution if unpackaged products are handled there. Where there is no choice about the location of the building, the orientation of the building or glazed facades can be selected to minimise pollution and over- or under-heating.

Apart from the possible direct effects of outdoor pollution, certain locations can create a need for mechanical ventilation and air conditioning, the latter being a major risk factor for SBS. Outdoor levels of noise and (mainly outside the UK) temperature may also therefore need to be considered.

4.2.4 Building services

(a) THE NEED FOR AIR CONDITIONING

While air conditioning can in many buildings be beneficial or even essential, it is right to question the circumstances under which complex ventilation systems are needed. In particular, questions about the need for cooling in temperate climates should be high on the agenda, and full consideration given to natural ventilation alternatives to air conditioning (see Perera et al 1988).

Where air conditioning is required, for example because of the requirements for equipment in the building security, or excessive pollution outdoors, the following must be considered.

(b) AIR SUPPLY

Air should be from a clean source, through a clean system, in sufficient volume, effectively filtered.

The requirement for clean air means that the intake should not be close to or downwind of vehicle access, parking, ventilation exhaust etc. In fact locating air intake upwind from exhaust is not adequate if only the prevailing wind is taken into account. Other wind directions can cause significant exhaust re-entry (Lepage & Schuyler 1990). Where outdoor pollution is a problem, buildings can be operated under positive pressure to reduce infiltration (Collett et al 1991) as is common in the UK. However in tall buildings the effect of warm air rising through the building means that infiltration is likely to occur.

Within the building, the location of pollution - generating activities and air flow need to be managed to prevent transfer of pollution. Effective filtration, designed for easy regular replacement and maintenance is also important. This is achieved by current good design, but not always by current practice.

The volume of air required for pollution dilution is probably adequately specified in current guidance (see 3.2) except for perhaps the first year of a building's operation. Ventilation can be increased 3 or 4 times with only 5% increase in energy costs (e.g. Ventresca 1991, Eto 1990) for example using a VAV system with an air economiser. However, if the capacity of a system has to be increased in order to cope with the initial period of ventilation this would probably not be cost-effective.

Bordass (1990) has argued that both the under-provision or over-provision of capacity in a system for heating, cooling or ventilating can lead to poor performance and thus to unsatisfactory environmental conditions. To avoid this situation, Bordass says "the best approach may be to meet 90 per cent of the expected range of needs, while making alterations and additions easy, and while providing the adaptability to accommodate longer-term change ... Such strategic provision must not be regarded as 'waste'. If it is not yet needed for services, then it can be used for something else."

Recirculation should be avoided if possible in favour of heat recovery, but the heat recovery mechanism is important. For example Schaeffler et al (1988) found that pollutants can be exchanged between indoor and outdoor air when air-to-air heat exchange wheels are used.

Fibrous materials in the air path, standing water and cold water humidification of the supply air can also be regarded as risk factors.

Antimicrobial coatings applied to the interior of HVAC systems can inhibit re-growth following cleaning (Rhodes & Gilyard 1990) but viable microbes may not be the cause of SBS, and it must be ensured that the coating itself has no adverse effects on the occupants.

Strindehag et al (1988) note that different humidifier types will result in different carry-over of bacteria from water to air, evaporative systems being better from this point of view. This, however, was a short term laboratory study which would not indicate the actual contamination possibilities with systems which have been in use for some time and, again there is very limited evidence that bacteria play a role in SBS.

Robertson et al (1990b) conclude that liquid desiccant-air conditioning systems can be attractive alternatives to refrigeration-air conditioning systems because the disinfectant properties of liquid solutions may improve IAQ by reducing airborne micro-organisms. However, to develop an efficient and economic system, the effects of parameters such as pH, temperature, concentration of desiccant solutions, and contact time need to be studied systematically. Solid desiccants have the capacity to remove chemical pollutants, but their capability to reduce airborne micro-organisms is not known. Several questions remain, including whether the solid desiccant surface will act to amplify the growth of micro-organisms once moisture is adsorbed.

(c) AIR DISTRIBUTION

Fresh air should be delivered so as to maximise the fresh air in the breathing zone of the occupants; it is wasted if it enters the building but does not reach the people. Air distribution and ventilation effectiveness (efficiency) are matters of current debate.

In particular the choice between dilution ventilation and displacement ventilation is the subject of polarised views. Dilution or mixing flow, as the name suggests, relies on a mixing effect of outside air with room air to keep the level of oxygen up and pollutants down. Displacement flow involves conveying cool air at low velocity into a room at near floor level. This layer of cool air will not mix with existing warm, polluted office air, but gradually rise as it warms, displacing the existing polluted air which can then be extracted at ceiling level. Pollutants generated below the cool air band but which are at a higher temperature can rise through this cool layer to be efficiently extracted (Andegiorgis 1988).

Displacement flow is now reported to be more popular (Fitzner 1987) because it saves energy, improves comfort and increases ventilation efficiency. However, Andegiorgis (1988) reports that if air disturbance is sufficient to break up the stable underlying layer of clean air, there will be an overall rise in the concentration of dust particles and gases. Palonen et al (1988) state that the resulting vertical temperature differences can lead to complaints of heat stress at head level and draught around the feet. Holmberg et al (1987) compared the two methods experimentally and found that as well as being less energy efficient, dilution flow with air supplied through wall mounted and floor mounted supply devices resulted in convection currents around a seated person which could result in a high concentration of tobacco smoke in the breathing zone.

Because of the complications involved in combining ventilation with heating and cooling functions, some (e.g. Cole & Rousseau 1990, Wyatt 1991) favour separation of these functions in order to optimise both. One approach is to use displacement ventilation as ventilation only (except that the air must be supplied at slightly less than room temperature in order to achieve separation of the layers of clean and polluted air) and chill beams to provide cooling.

Jackman (1991) provides a brief description of displacement ventilation systems, but the verdict probably has to be open for the moment: they can operate effectively, particularly in the summer but it remains to be seen whether the technology (and knowledge of the technology) is adequately developed to provide good conditions with higher probability than dilution systems.

(d) OCCUPANT CONTROL

One of the more difficult problems is how to achieve good ventilation system efficiency (and low energy costs) while providing a degree of user control, particularly in existing buildings. However, Drake et al (1991a,b) suggest on the basis of case studies that user-based control systems are not only feasible but advisable. A single set of environmental conditions will not suit everyone, and low individual control is associated with annoyance, stress, high symptoms and low productivity (Raw et al 1990). It is not clear whether this is because occupants achieve an indoor environment more suited to their needs or because they are more tolerant of poor conditions if they feel they are in control. Individual control can also in principle afford greater energy efficiency (not necessarily economy) because energy goes where it is needed.

The simplest approach to control of ventilation and heating is to have openable windows (with trickle facility) and radiators with thermostatic valves. This can be enhanced by providing computer-generated signals to advise opening and closing. This provides limited cooling on the hottest days of the year in the UK. Where mechanical ventilation is used, a variety of user-based control systems are available, some meeting only user group requirements and some the requirements of individual users through individual workstation control. Openable windows are commonly felt to be incompatible with air conditioning, but Bordass (1989) has demonstrated that a building can feasibly combine openable windows and air conditioning to produce a 'mixed-mode' solution without energy penalties.

Lighting controls vary from individual controls for task lights or uplighters to zone and floor groupings of luminaires. Currently, grouped arrangements can be controlled from individual or banked switches, or in association with automatic lighting control systems. The latter permits the grouping of luminaires to be changed at will, without rewiring, and can be programmed to operate on a predetermined time pattern with overrides from daylight levels or an individual's needs. Like temperature control it too can be coupled to a building management system.

Any approach is likely to be enhanced by user education and involvement to reduce unintentional abuse, give users a greater perception of control, provide information for managers and diffuse complaints, particularly if the user must access an computer system (e.g. by telephone) in order to exercise control.

4.2.5 Building and furnishing materials

There is incomplete knowledge about the role of building and furnishing materials, as there is for all the potential causes of SBS. In this situation there is always the (economic) danger that any action taken could be the wrong action. However, many organisations are now drawing up specifications and guidance for new offices in order to reduce the risk of SBS. There are too many different schemes to describe them all, and some are still under development but one scheme is included here as a detailed example.

The State of Washington requires that its new office buildings, currently in the design/build and construction process, "offer a productive as well as a safe and comfortable environment for its occupants" (Black et al 1991a). In order to achieve this goal, an assessment and implementation programme for indoor air quality has been designed. The major impact of the scheme is likely to be on building and furnishing materials, although it also incorporates building location, design and operational controls. The programme is intended to continue through construction, occupancy, and operation of the building, including training of building occupants and support personnel as HVAC technicians.

A dual approach is taken to minimising pollutant emissions: specification of materials and a building 'flush-out' prior to occupancy. The latter is discussed in Section 4.3. The scheme limits pollutant contribution per product or activity to maximum values of 500 $\mu g/m^3$ of TVOC; 50 $\mu g/m^3$ of respirable particles; and 60 $\mu g/m^3$ of formaldehyde. Using environmental chambers and computer exposure models, prediction of resultant chemical emissions from materials such as office furniture, adhesives, flooring systems, wall-covering systems, office machines, and other sources are computed based on the actual building design and operational requirements. Manufacturers of interior furnishings and materials are required to submit pollutant emissions data with their bid, indicating their compliance with actual specifications established by the IAQ programme.

Those individual compounds identified in the product emissions that meet the following requirements must be individually reported along with their predicted building concentrations:

- those contained on the list of chemical carcinogens established by the International Agency for Research;
- those contained on the carcinogen list of the National Toxicology Program;
- those contained on the Reproductive Toxin list as included in the Catalogue of Teratogenic Agents;
- those more than 1/10th the threshold limit value according to the American Conference of Governmental and Industrial Hygienists;
- those on the National Ambient Air Quality Standard Lists of primary and secondary air pollutants.

These criteria are said to provide a safety margin over current occupational chemical exposure standards and those currently referenced for indoor air (WHO 1987, Mølhave 1985, Tucker 1986, Otto et al 1990). A modified IAQ model (Sparks et al 1989) is used to predict indoor air concentrations of the pollutants of concern. It is not clear, however, how the criteria relate to the concentrations of pollutants which will be found in the air of the building.

It is interesting that, of the results of testing reported, all four furniture manufacturers complied with the pollutant emission criteria. Does this mean that the tests were not worth doing, or that the manufacturers changed their products in response to the State requirements, or that only the manufacturers who were confident of their products bothered to apply? It would be worth knowing the answer before proceeding with the scheme or introducing it elsewhere. Certainly it is possible to have not only furniture with very low formaldehyde and particulate emissions (Strobridge & Black 1991) but carpets and adhesives with low VOC emissions (Davidson et al 1991).

As a result of the development of this scheme, Black (1990) lists many product-specific factors affecting TVOC emission rates: product history, unexposed age, exposed age, packaging, storage, conditioning, homogeneity, inter-intra manufacturing variations, loading factor, loading form, sink/re-emission factors, environmental factors and quality control. All these can be considered when specifying materials. The US EPA is producing a catalogue of building materials which are sources of indoor air pollution (Stockton et al 1991). Saarela & Sandell (1991) are working to establish Nordic testing and guidelines for emission of VOCs from flooring materials, with 3 bands of emission rates with associated ventilation rate requirements.

Some useful information on the selection of building materials with regard to their toxic and irritant effects is provided by Curwell & March (1987). The Construction Products Directive will in the future increase the amount of information provided by manufacturers in the UK.

A number of programs for modelling indoor contaminant levels have been produced and reported for example by Kim et al (1990), Creuzevault et al (1990), Guo et al (1990), Haghighat et al (1990), Sparks & Tucker (1990) and Weir et al (1990). These would all have their limitations but they are very useful when dealing with multiple sources of pollutants and varying indoor climates.

Some are already taking independent action on what is already known, for example Fredericksen (1990) seeks to specify buildings with good IAQ by selection of low emission easy-to-clean materials. His clients are apparently satisfied but no objective evaluation is reported.

Selection of materials on the basis of an odour assessment may become possible as a result of the work initiated by Fanger et al (1988). This could reduce the cost of testing but the technique itself is not yet fully proven and would in any case not be applicable to non-odorous pollutants. It remains an option for materials specifiers to judge for themselves whether a material is odorous, but uncontrolled testing conditions could lead to unreliable results.

All this is based simply on the notion that it is generally good to reduce indoor pollution. There is no established target level of pollution which would necessarily eliminate SBS. Thus, although Building Regulations could be applied to require 'proper materials' it would be difficult to state criteria in relation to SBS. The availability and awareness of less polluting materials is of greatest importance and it is to this that attention should be given.

4.3 THE CONSTRUCTION PROCESS: BUILDING, INSTALLATION AND COMMISSIONING

4.3.1 Summary

Building, installation of equipment and services, and final commissioning of the building should all follow the design as precisely as possible. Any changes which are made to the design need to be checked to ensure that the building as a whole will still perform as intended (or better). For example, HVAC systems should be installed for ease of maintenance and properly commissioned whatever the pressures of time. Materials should not be substituted without consideration of the consequences for emission of pollutants. Consideration can be given to 'baking out' the building to drive off volatile pollutants.

4.3.2 Action during construction

It is perhaps obvious, but nevertheless needs to be stated, that building, installation and commissioning should follow the design. By implication, the design should specify the construction process where necessary. The correct installation and commissioning of HVAC systems is particularly important and this stage of the construction process is most vulnerable to pressures of time and finance if not carefully monitored by clients.

One thing in particular which has been known to result in problems is installation with insufficient consideration of maintenance. The design of systems has changed over time to make them more amenable to easy maintenance but this is wasted if they are installed in inaccessible positions. But other gross errors have been known to occur: ductwork missing or not connected, 'moving parts' which do not move, fans omitted or reversed etc.

If particular building and furnishing materials are specified for reasons of low emission of pollutants it is also important to ensure that there is no unauthorised substitution of materials. Care must also be taken to remove any volatile chemicals which are spilled on or near the site and to protect the integrity of any barriers which are designed to contain matter such as fibrous insulation. See Curwell et al (1990) for more detailed discussion of risks during the construction stage.

Failure to commission building services correctly can obviously cause problems, and not only SBS, but this is mainly a matter of applying established knowledge and techniques according to instruction, albeit in a complex field. This is therefore dealt with primarily as a management issue (section 4.6.3).

There are certain aspects of good practice which are sometimes forgotten at the final stages of construction which, if included, could reduce the subsequent indoor pollution. It is all too obvious (but sometimes neglected) that all parts of the building and services should be free of excess moisture, dirt, dust, litter and other extraneous material when the building is handed over. More recently the practice of 'bake out' has been the subject of debate.

4.3.3 Bake-out

Bake-out is the term used to describe the process of heating and ventilating a building to drive pollutants (primarily VOCs) off materials before a building is occupied (or as a later remedial measure). Without the heating, the process is sometimes referred to as flush-out. There is debate over three main issues: does bake-out work, is it cost-effective and does it damage the building? The evidence is mixed.

Girman et al (1990) state the problem well: bake-out must be conducted within constraints of cost and potential for damage to the building (e.g. wood shrinkage, metal expansion, cracking or peeling of finishes, reaction of heat-sensitive sprinklers). But sufficient temperature and time must be allowed to raise the temperature of shielded materials such as carpet glues. They conducted five case studies and showed that bake-out can reduce VOCs at low cost, around £1-2 per m^2 at 32°C and normal ventilation rates over 2-3 days. A few defective components were damaged but otherwise the bake-out was conducted safely. Regular walk-through inspections, and the removal of items likely to be affected by heat, are useful precautions.

Other studies have shown less positive results. Bayer (1990), in a laboratory trial of bake-out of particleboard and office partitions showed little reduction in emissions, and some increases. Hicks et al (1990) conducted a bake-out at 38 - 43°C; there was no damage but little reduction in VOCs and an increase in some. However the ventilation rate in this study was only 0.2 air changes per hour so there may have been transfer from relatively sealed sources such as laminated particleboard to open pollution sinks.

In Washington State, an initial requirement for a 90-day 'flush-out' period with 100% outside air was changed following early experience to allow a 30-day period prior to initiating furniture installation (Black et al 1991a). This decision was made to prevent VOC contamination of the furniture by the adhesive and other 'wet' pollutant emitters through potential sink effects. An alternative or complementary approach would be to require a certain period of storage for materials in a well ventilated space before delivery to the building.

Further work is clearly needed before bake-out specifications can be drawn up with confidence.

4.4 EXISTING BUILDINGS: OPERATION AND MAINTENANCE

4.4.1 Summary

Design, construction, installation and commissioning should all be carried out with maintenance of the building and its services in mind. Maintenance must then be carried out according to schedule produced in the course of design and commissioning. Maintenance should include attention to hygiene in ventilation systems and in relation to office cleaning, repair and adjustment of building services. All this requires good documentation of the building services and access to the components which need to be maintained. The pattern of heating, cooling and ventilating can be used to minimise airborne pollution during the occupied period.

4.4.2 The need for maintenance

That buildings in the UK suffer from poor standards of maintenance has been acknowledged by various professional groups in their evidence to the Environment Committee inquiry into Indoor Pollution (1991). The problem is also noted by the Second Badenoch Report (1987). The HVAC Association provided the Environment Committee with a catalogue of potential maintenance problems. Major reasons for poor maintenance are the low status and inadequate training of many maintenance personnel and poor procedural control of maintenance operations.

Buildings and their services have grown more complex over the years, while the need for suitably trained maintenance staff has not been fully understood or acknowledged. Consequently there are not sufficient trained personnel with the necessary skills to maintain the growing number of such buildings, so that the environmental standards at handover (assuming that the building services have been commissioned to meet the design intent) are achieved over the life of the building.

Responsibility for this must be shared among several sectors of the building industry. Developers often let buildings on a full repairing lease, relieving themselves of maintenance functions. Designers do not always emphasise sufficiently the importance and cost benefits of good maintenance. Estate managers may not appreciate the importance of good maintenance, or simply pass on any service and repair charges as part of the landlord's costs. Occupiers may also be unaware of the importance and cost benefits of good maintenance and as a result may suffer loss of productivity through poor working conditions and/or SBS. The building services industry has not provided adequate training courses, or marketed 'good maintenance' concepts well enough.

This Section is not intended to be a comprehensive guide to maintenance practice. However, it is worth highlighting those operation and maintenance activities which are most likely to improve the features or factors most frequently suspected of contributing to SBS. In general, all parts of any HVAC system should be maintained to the state of being operational at all times.

4.4.3 Hygiene

Continuous vigilance is needed because the presence of so much organic material is a potential danger area for microbial growth. This, in terms of building services generally, means regular cleaning of all heating and cooling terminals, constant attention to debris and dust which collects on pipework and ductwork and rigid adherence to all filter servicing and replacement requirements.

Occasional internal examination of ductwork will act as a check on filter performance and indicate whether a cleaning procedure is required. However good the quality of filtration, dust is created within the modern office which is not removed. It is therefore necessary to allow for regular, but hopefully infrequent, cleaning above false ceilings, below false floors and in ductwork.

Water treatment is a complex maintenance problem where specialist advice is needed, but in air-conditioned buildings it is important (there may be no such requirements in other buildings). For open water circuits associated with cooling towers the correct treatment is crucial to ensure, amongst other things, that Legionella is not disseminated, but this is not a problem with SBS. Any treatment applied to water systems which may related to SBS will be to humidifiers, sprays and washers in air handling units providing air to occupied spaces. Where such treatment is considered essential, it needs to be ensured that no noxious or toxic elements are present which could be carried over into the space by air being passed over or through the treated water.

The other major requirement for hygiene is in fabric furnishings and storage areas where dust can gather. Wherever possible an excess of such materials and storage areas should be avoided. Occasional treatments to remove mites (e.g. steam cleaning, acaricides, liquid nitrogen) may be beneficial. Normal vacuum cleaning is quite ineffective for this purpose (Bischoff & van Bronswijk 1986) whereas acaricides can be very effective, leaving the allergenic excreta to be removed by subsequent cleaning (Bischoff et al 1990). However the case for implicating dust mites in SBS is not yet fully proven and it would be advisable to identify mites in the furnishings (particularly chair covers) before proceeding with costly eradication measures.

4.4.4 Heating, ventilation and air-conditioning

In air-conditioned buildings, the volume of fresh air should be checked regularly (particularly where there are changes in heat load in the building which might create anomalies in the recirculation rate) and this must encompass the proper setting of all damper arrangements associated with fresh air, recirculated air and exhaust air ducts. The design minimum fresh air quantity must always be available during occupation.

Regular checks at a representative random sample of locations should be included in maintenance routines to check air distribution and velocities at work stations and ensure that they are kept within the specified limits. Random sampling of the calibration of space temperature sensors is important and any errors should be corrected. Regular routine maintenance of all such sensors, probably biannually, should be included - possibly through a contract with the equipment manufacturer.

Humidity controls should be frequently checked and included in the regular maintenance programme - as should the complete control system. The frequent checks should include the operation and cleanliness of the actual humidifying equipment/system.

4.4.5 Lighting

Regular maintenance, including bulk replacement of light sources at the end of their useful lives, is important to maintain a comfortable working environment. There will be intermittent failures of the odd lamp or luminaire between regular maintenance visits and these should be dealt with without delay. Switching systems should be regularly checked and those associated with automatic switching should have their operational programme verified to see that it suits any changes in occupancy patterns.

4.4.6 Information and access

The maintenance areas mentioned above should, of course, form part of an overall regime which evolves during the design and installation process and is finally included in the Operating and Maintenance (O&M) manuals which have to be made available to the developer/occupier/managing agent. The O&M manuals cannot be finalised until the 'as fitted' drawings have been produced after practical completion, but comprehensive drafts must be available before commissioning commences.

The Health and Safety at Work Act (1974) requires a person who designs, manufactures, imports or supplies any article for use at work to provide adequate information about the use of the article, including any conditions necessary to ensure that it is safe and without risks to health. This applies to HVAC systems and the information is in this case most usefully included in operating the maintenance manuals. It is not necessarily an offence not to provide the manuals by the time occupation takes place, but it is certainly desirable.

Maintenance, however well-planned, defined and documented can only be properly implemented if the designers recognise the spatial and access requirements for such operations, i.e. access to ducts, filters and other items of building services plant. This is covered by a general requirement for access under the Health and Safety at Work Act (Section 2) but this legal requirement may not be clear to all designers or installers.

Access space must be provided generously at the planning stage since, once the building is in operation, additional plant and equipment may be installed which will inevitably reduce accessibility for maintenance. Major items of plant need to be positioned with sufficient space to enable servicing, maintenance and replacement of parts, which may need long withdrawal zones. The equipment also has to be configured so that instrumentation can be seen, read and utilised.

Access to all equipment mounted around the perimeter, above false ceilings and below false floors is essential. Maintenance requires regular access to all such equipment and inadequate, poorly detailed arrangements will have one of two effects:

- the maintenance will not be carried out;
- it may be carried out with difficulty and therefore not regularly, leaving as an aftermath ruined ceilings, damaged perimeter decoration and uneven floors.

4.4.7 Operation to minimise pollutants

Nielsen (1988), based on a theoretical analysis of emission, adsorption and re-emission of pollutants, makes the following operational recommendations for avoiding problems:

- continue heating and ventilating for a short period after the end of the working day;
- keep temperatures as low as is comfortable in occupied rooms;
- avoid temperature rises during the day (assuming the temperature is adequate at the start of the day);
- heat the building to 5°C above intended operating temperature before the occupants arrive, then cool to operating temperature using 100% outside air;
- use high air velocities to remove pollutants from fabric surfaces when the rooms are not occupied;
- institute effective cleaning in offices.

4.5 EXISTING BUILDINGS: SOLVING PROBLEMS

4.5.1 Summary

Clear diagnosis of SBS should be the first step, and this will normally involve a questionnaire survey or medical interviews, possibly enhanced by spatial analysis of the distribution of symptoms in the building and corrections for workforce characteristics. If such a survey is not possible, environmental monitoring will have to suffice but it will be difficult using this approach to demonstrate a satisfactory outcome, since there are not definitive criterion levels.

Attempts to cure cases of SBS should start from the dual premise that many factors may be interacting to cause the problems and that things are not always as they seem. There may be obvious faults in a building, and the occupants may attribute their symptoms to these faults, but the faults and SBS may or may not be related. Nevertheless, any obvious faults should be corrected before embarking on extensive monitoring surveys.

Decisions about treatment should consider all the options, with reference to all the known risk factors and established ways of dealing with them. Where possible, the success of treatments should be verified by controlled before and after surveys.

4.5.2 Introduction

As stated above, prevention is better than cure, but we probably do not have the knowledge to achieve 100% prevention and there are plenty of existing cases of SBS. In dealing with these cases, two types of approach are possible (as extreme examples): an attempt to reduce complaints as quickly and as simply as possible, and a research study to solve the problem and say how it was solved. Short-term economic constraints mean that the former approach is the more common; this is understandable but unfortunate because such investigations provide limited evidence about how to deal with future problems in the most cost-effective manner.

Whatever the approach taken, it is important to remember that there are many possible causes of SBS and they are interrelated and interactive. Except in a few cases where symptoms follow a specific change in the working environment, e.g. the installation of a new carpet, isolation of the cause is notoriously difficult. SBS is a multifactorial problem which demands a multidisciplinary approach: a comprehensive view and systematic checking of possible problems, not a standard approach applied to all buildings. Only then can building owners and building users be assured that problems are being effectively dealt with.

Any investigation must comprise two components: diagnosis and recommended cure. A confident (or possible unwise) consultant will offer prognosis. The responsible consultant will offer valid follow-up testing of the outcome of interventions, although such offers will not always be accepted by clients. Follow-up testing does not have to be done in the whole building but could be used in a 'pilot' of a remedy in part of the building.

4.5.3 Diagnosis

In many investigations, the occupants may consider they have already made a diagnosis. Such feelings are easy to understand but they may be misleading. Unless spontaneous complaints are investigated in a more formal manner - to determine, for example, whether it is the building or the management the staff dislike - a lot of wasted effort may go into the cure.

Of course in some cases the occupants may not wish to have a precise diagnosis - just some quick action. A trained eye may quickly identify some obvious problems with a building, for example a contaminated humidifier or dampers seized in a position which prevents outside air from entering a building. There is no guarantee that putting these things right will eradicate SBS, but they clearly need to be put right anyway. Beyond such obvious faults, the investigation becomes more like a research project and reliable independent consultants should be considered.

A questionnaire survey (using an established questionnaire design) of the building occupants is the standard diagnostic tool, although an interview survey is a more costly but equally valid option. Either can be valuable as a preliminary to remedial measures, to establish the nature and number of the complaints, whether the complaints come from throughout the building or only certain parts, and what may be causing them. It will also provide a means of determining whether any subsequent remedial measures have been effective.

An occupant survey is clearly not possible unless the building is currently occupied and even then the organisation occupying the building may not agree to such a survey. If a survey is feasible, then it should be carried out by an organisation which can compare the results with a large sample of other buildings in order to indicate the relative level of SBS. Of course not all the causes arise directly from the building. Even a questionnaire survey may therefore not be valid if there is in the future a change in the individuals or organisation occupying the building.

A possible enhancement of the standard questionnaire approach is reported by Hedge & Ellis (1990): spatial analysis of symptoms and environmental conditions within a building. This has some potential in large buildings but must be used with great care because the output could be governed by the location of a few sensitive individuals, or by age/gender/job being associated with location. In any case, as the authors are aware, more accurate diagnosis can be obtained by correcting for the job type and gender profile of the workforce in a particular building.

In theory it would be possible to use a panel of independent observers to occupy a building for a time and report their symptoms. Unfortunately they might have to spend 6-12 months in the building before giving an accurate report. Using sensitive individuals ('human canaries') who would react more quickly might overcome the time lag problems but there would be ethical difficulties and Grande & Hanssen (1988) largely discount this option because allergic people's reactions are governed by too wide a set of circumstances, not just the indoor environment.

An apparent alternative to a questionnaire survey is to carry out environmental monitoring in the building. For example Gammage et al (1989) offer a number of empirical guideline figures for screening purposes. These include 1000 parts per million of carbon dioxide to screen for outdoor air supply, 500 colony forming units per cubic metre of micro-organisms to screen for sanitary conditions and 1 part per million of volatile organic compounds to screen for active sources of these chemicals. This approach may of course indicate parameters which are outside acceptable ranges, but it cannot strictly speaking be used to diagnose SBS and may be very wasteful if it turns out that complaints are localised within the building or occur only in certain occupational groups or at certain times of day or week. However, where it is felt that occupants may not provide honest responses to a questionnaire the monitoring approach may have to be taken.

Similarly a panel of observers can be used to assess the environment, as in studies of odour, but again a clear relation between these assessments and SBS has not been established.

4.5.4 Cure

(a) GENERAL GUIDANCE

There is only limited empirical evidence about 'cures' and the complex nature of SBS. Taking the optimistic view, it is very likely that among the hundreds of (scientifically speaking) unsubstantiated claims of cures, there is a significant number of genuine cases. Certainly many malfunctioning, ill-maintained and polluted buildings have been improved. This experience, together with a good understanding of the workings of buildings, can be used to make recommendations.

A very helpful guide from the Danish Building Research Institute (Valbjørn et al 1990a) does just this, and its recommendations are summarised in Appendix 2;

there are other guides available, e.g. Public Works Canada (1990), de Roo (1988) or Davidge & Kerr (1990) but this is probably the most comprehensive and it is simply presented. Where there is obvious discomfort due to lighting, noise or static shocks, action should also be taken to reduce these problems.

Most of the recommended remedial measures are based on well-established knowledge but it is worth discussing some specific approaches to indoor air quality problems; there are effectively four approaches (Sparks 1991):

- increased ventilation, possibly outside the period the building is occupied (best for multiple low-level widespread sources);
- more effective (e.g. local) ventilation (good for localised, particularly intermittent, sources if the location can be isolated);
- use of air cleaners (best for larger particulates but need regular maintenance);
- source management - removal (e.g. smoking ban), isolation or replacement (i.e. with material which has a lower emission rate) - the best approach where it is feasible.

Options should in general be selected to minimise deposition of pollutants in sinks (surfaces from which they can be re-emitted), and where possible to replace materials (e.g. textiles, paper) which can act as sinks: sinks mean a pollutant has to be managed twice.

In practice the first point of action is often the building services engineer, in which case a sequence of priority actions can be suggested:

- ensure the building services are clean;
- check for failure of regular maintenance items and malfunctioning parts;
- adjust operation or modify the system as appropriate, particularly if there have been changes in demand or layout in the occupied space.

These actions may not solve all the problems but they are appropriate as a first approach. Beyond this, observations of possible sources of pollution should be made (including materials, equipment and occupants, and in older buildings, the cleaning regime).

If these initial investigations do not produce a satisfactory outcome, more detailed monitoring and surveys will be required.

Where specific evidence is available on the effectiveness of any measures, it should of course be used as a guide to action. For example, Kerr & Sauer (1990) tested three methods of reducing VOC pollution from liquid process photocopiers: carbon filters were not adequate and dilution ventilation was adequate for health, but probably not comfort. Local extract ventilation is usually recommended for this situation.

Most of these approaches have already been discussed but air cleaners are a current issue which can be discussed briefly here.

(b) MECHANICAL AIR CLEANERS

Air cleaners are of two principal types - those which remove particulates (e.g. by filters or electrostatic precipitation) and those which remove gases and vapours (normally using activated carbon). These are discussed in a little more detail in Indoor Air Quality Update (August 1991). Particulate filters are relatively well-established and Brown et al (1991) provide evidence that activated carbon filters can reduce VOC levels. Yu & Raber (1990, 1991) present calculations to show how particulate and gas phase filters can be used as an alternative to raising ventilation rates by the 50% implied by the results of Fanger (1990).

A relatively new method of air cleaning is to use ozone to convert VOCs to water and carbon dioxide. Rodberg et al (1991) describe such a device which is built into HVAC systems and uses a final ozone destruction catalyst to remove residual ozone. It was found to be 50% effective under the conditions studied, with no residual ozone. As Shaughnessy & Oatman (1991) point out, use of ozone as an air cleaner within occupied spaces is not advisable and any ozone system should be designed, operated and maintained to prevent ozone being released in the occupied space.

(c) GREEN PLANTS

For several years, the US National Aeronautics and Space Administration (NASA) has been investigating the ability of plant leaves and roots to remove trace levels of toxic chemicals from closed experimental chambers (Wolverton et al 1984, 1985, 1989a,b) NASA's concern is the potential air pollution problems associated with closed structures in outer space, but the research has relevance to removing pollution from buildings and indeed is already being applied in the UK for this purpose.

The experiments examined the ability of a range of foliage plants to absorb carbon monoxide, nitrogen dioxide, formaldehyde, trichloroethylene and benzene at a range of initial levels, with and without the aid of fan-assisted filtration by activated carbon in the growing medium. Plants can also be used to increase humidity and remove carbon dioxide, even over night (Raza et al 1990).

These experiments have shown that plants hold great promise for the improvement of air in buildings but it has to be emphasised that all the studies were short-term laboratory experiments and there have been no reports on longer-term field trials. The results do however suggest that, if enough are used, the passive use of plants could be sufficient to maintain clean air. However, plants grown with their roots in activated carbon, through which air is drawn, remove airborne pollutants more rapidly and efficiently. What is particularly interesting is that the limited longitudinal data available indicate that the effectiveness of an individual plant to absorb and metabolise a chemical increased when it was exposed to the chemical over a long period of time. This may be due to the pollutant inducing higher activities in the metabolic pathways involved.

Although some species were tested for the emission of potentially harmful metabolites, and levels were negligible, this should perhaps be investigated further. The report did not specify which species were used or which substances were tested for. It was stated that the experiments were performed under low light intensity, and that as temperature and light levels are increased, there will be an increased rate of pollutant removal, but also more metabolites will be given off. The presentation of the results for the removal of low concentrations of trichloroethylene, benzene and formaldehyde is unsatisfactory because no information was given about the size of the plants used or their relative efficiency.

When large numbers of plants are provided under contract, the suppliers will usually replace diseased or infested specimens. If plants have to be treated in situ, systemic pesticides might be safer than sprays. The use of biological controls might also be considered in tightly constructed buildings. These can take the form of parasitic insects or fungi which specifically attack plant pests.

It is possible that micro-organisms could grow on moist surfaces and cause a hazard to human health. For example, various species of Aspergillus fungus can colonise the soil of potted plants, and their spores are capable of causing allergic reactions and, more rarely, lung infections in susceptible individuals. The reality of this risk should be assessed in order to provide assurance that plants do not introduce new problems into buildings.

(d) MAJOR REORGANISATION

There are clearly some things which are very difficulty to improve by retrospective measures, as distinct from design. In particular, creating cellular office layouts from open plan spaces can be difficult because of requirements for air distribution, and achieving occupant control of the environment is a complex process, especially if the fabric of the building and the building services are poor to start with. In such cases, formal cost-benefit calculations may be needed to evaluate extreme options such as abandoning the building - temporarily or permanently.

4.6 THE ROLE OF MANAGEMENT

4.6.1 Summary

Good management can avoid the environmental problems and workforce stress associated with SBS and ameliorate adverse staff reaction to problems which do occur. This will demand that SBS is recognised as a problem which is both important and potentially soluble, not just an excuse for complaint. On a wider scale, management action throughout the life of a building may require that management recognise the value of a building as a corporate and economic asset which can have a positive effect on both the workforce and the company image.

But the motivation to act is of little value (and may even cause problems) unless management has the capacity to take appropriate action from the design brief through to the occupation and operation of the building. In particular the critical role of commissioning and maintenance need to be translated into a demand that these be designed in and executed according to design. Even if the systems are 'right', problems will arise if the complexity of the systems is not matched by quality and quantity of management input to operation and maintenance, including monitoring and complaint management.

4.6.2 Definition of the problem

Among the factors which are commonly cited as causes of SBS is the 'quality of management'. Little elaboration has been provided on the nature of the management factor but Dawson et al (1988) have defined two essential conditions which are required of management in dealing will health and safety issues: management must have 'the willingness to act' and 'the capacity to act'. These two aspects of the management problem are discussed below. It is also of course necessary that management should have the knowledge required to take the correct action, whether this depends on their personal knowledge or advice given by others.

4.6.3 The Willingness To Act

There are two reasons why managers might be unwilling to act in ways which will prevent or control the incidence of SBS. The first reason is that operational managers simply do not recognise the problem of SBS, the second is that buildings as a whole are not valued by senior business managers.

(a) RECOGNISING THE PROBLEM

SBS is still a controversial condition and early conferences on the subject of SBS (e.g. CIBSE 1987) indicated the level of scepticism surrounding the syndrome with titles such as 'SBS - myth or reality'. The scepticism derives from the fact that SBS probably has no single environmental cause and also from the fact that the symptoms are not readily measurable by objective methods but rely almost totally on self-report. This latter criticism has been dealt with (see Section 3.8.2).

However, despite support for the validity of self-reported symptoms, the lack of a single physical cause remains a problem. Complex conditions such as SBS, with multiple causes, can be difficult to understand and this is a deterrent to taking action. Also problematic is the fact that psychological factors have been described as playing a role in the causation of SBS. This is often misunderstood by managers who attribute the problem to individual psychological pathologies: that is, to a group of habitual 'moaners', rather than to negative reactions to poor quality environments. In dismissing the problem in this way, managers are failing to discriminate between valid and invalid complaints; valid complaints are thus ignored, and individual sensitivity is likely to be increased.

Another factor which may inhibit willingness to act is perceived capacity to act. For example one factor which may contribute to SBS is the spatial design of the building. Deep spaces with low floor-to-ceiling heights are perceived as claustrophobic and alienating, especially if the building has a sealed skin (Tong 1991). These are inherent features of the building design and so are not subject to management intervention, and this obviously reduces managers' willingness to act since they perceive the entire problem to be beyond their control. Failures to solve problems by environmental improvements (e.g. Boyce 1989) tend to confirm this belief: Boyce made remedial but non-structural alterations to a deep-plan, sealed building with high rates of SBS symptoms and yet succeeded in reducing the symptom rates only marginally.

It is clear that managers who are faced with a building which appears resistant to improvement have limited options. However, managers in this situation must ensure that they do not increase the sensitivity of staff by ignoring the problem or being slow to act on obvious defects which they can remedy. It is crucial that managers do not become 'unwilling to act' or fall into the belief that they can do absolutely nothing; the importance of social and psychological factors in SBS have been mentioned above, and managers who are genuinely concerned about the problems suffered by their staff can often compensate for fixed environmental deficiencies by adopting other strategies for improvement of people's working environment.

For example, a naturally ventilated building repeatedly overheated in summer and management felt they could do nothing. Tong (1991) suggested pre-cooling of the building prior to the working day, the removal of screens which were obstructing air flow, and the removal of clutter from in front of some of the windows to make them easier to open. These measures, although they only achieved a marginal improvement in conditions, were found to be effective in making staff feel more satisfied. In situations like this, one approach is to consult the staff about possible workplace improvements. Staff may be aware of opportunities for improving workplace quality that have eluded management. This type of consultation must be done professionally and with care to avoid producing a 'wish-list' which is impossible to fulfil, thus frustrating staff even further. It is also important to be aware that such 'proxy solutions' may be only temporary if the original problem persists.

(b) RECOGNISING THE VALUE OF BUILDINGS

Unwillingness to act can be manifest both amongst technical managers such as facilities managers and maintenance engineers, and amongst senior organisational managers - "Top managers tend to deal with facilities issues in a very discontinuous fashion. Workplace environmental issues, when raised by others in the organisation, are dismissed as trivial, peripheral or irrelevant to the 'real' business decisions" (Steele 1986).

Further evidence of this is provided by Wilson (1985), who examined management attitudes to buildings and developed a hierarchy of building value and described five categories of role which managers believe buildings perform:

- the container role: the building as an organisational container, a necessary, if regrettable, overhead which should involve the minimum of expenditure;
- the public relations role: the building as a corporate symbol of prestige, with the emphasis on dominant location and impressive exterior image;
- the industrial relations role: the building as a tool of personnel relations, with interior standards of comfort and safety conveying concern for the workforce;
- the instrument role: the building as an instrument of efficiency, with layouts and services consciously designed to aid productivity and support internal communications;
- the inspirational role: the building as an inspirational force with expresses the high standards of attainment sought of all staff and, particular, a commitment to excellence.

Clearly, business managers who regard buildings merely as 'containers' will be less likely to place a high value on environmental conditions, thus possibly making these buildings more prone to SBS.

Even in buildings which are managed so as to be an instrument of efficiency, the environment can lack qualities which are increasingly being sought by occupants (Tong 1989). The emphasis on efficiency can sometimes be detrimental to 'total building quality'. Tong compares this with underpasses: "The underpass is intended to be an efficient way of enabling people to cross the road quickly without risk of accident". Yet underpasses can be "inhospitable, alienating and in fact, dangerous". The buildings on a major business park have been described as "a product of mechanistic minds apparently unaware of the importance of human emotions that need satisfaction" (Hodgkinson 1991).

The driving force in many building designs has been maximum floor area, as a basis for offering flexible layout options and easily-adaptable space planning arrangements. The resulting deep spaces have been combined with high specification environmental services to create some quite unattractive environments. The way of thinking which has led developers to focus on the instrumental role of buildings has come under attack as a result of "the realisation among the most sophisticated British developers that consumers of office space are becoming more demanding and .... the growing awareness that an alternative ..... in many ways extremely attractive way of providing office space is flourishing on the other side of the English Channel" (Duffy 1989).

Duffy outlines the more humanistic paradigms for office development which are evident in Europe:

- office as college - Colonia Insurance Company, Cologne;
- office as village - NMB Bank, Amsterdam;
- office as the place of creativity - SAS Headquarters, Stockholm.

It appears that Europeans are not simply more keen to maximise the 'inspirational role' of buildings through improvements to the design of the built product but also want to develop the 'industrial relations role' of buildings through improvements to the design process. The chairman of the German Publishers Gruner & Jahr who have recently built a new headquarters building, has said that "one of the principles of modern management is, are far as possible, to consider the needs and wishes of individual workers in the design of their workplaces" (Schulte-Hillen 1991), and so discussion groups were set up to involve building users in the design of the headquarters.

Such examples of enlightened management action in building design are not unknown in the UK. The development of the National Farmers' Union headquarters is one example where a successful design has followed user involvement. However, this approach is more common in Europe because the tradition of building development is one of custom built, owner occupied buildings whereas the British tradition is a "vendor-driven and landlord-let system" (Duffy 1989).

Can European ideas can be absorbed into British office development? One argument which is likely to encourage UK business managers to take their buildings seriously and to press for higher standards is the evidence that SBS has an economic impact on the company which occupies the building (see Chapter 2). This evidence means that SBS should be recognised as a problem for personnel and business managers rather than being treated as a remote, technical issue which is only of relevance to facilities and premises managers.

Buildings would then be seen as not only as functional business tools but as a means of enhancing staff well-being and performance. This would encourage a focus on both the industrial relations and the inspirational role of buildings, helping to forge a closer relationship between development and the aspirations of business managers. To date, SBS has not been presented in this way.

4.6.4 Capacity To Act

(a) WHAT IS THE CAPACITY TO ACT?

According to Dawson et al (1988) the term 'capacity to act' covers "what is involved at a technical level if hazards are to be identified and controlled". The following paragraphs therefore deal with the appropriate systems and procedures which are needed to manage buildings at both the strategic and the tactical level. The systems and procedures which are relevant to the prevention and control of SBS can be examined in relation to the life cycle of a buildings, as in the earlier parts of this chapter.

(b) DESIGN AND CONSTRUCTION

Managers can intervene in the design process by influencing the development of a building brief. The brief, and monitoring of construction, must address specific issues relevant to the control of SBS, as described earlier in this chapter. At the tactical level management can also include staff in the briefing process: note the role of staff consultation and participation in the EEC Directive 89/391 which deals with the introduction of measures to encourage improvements in the safety and health of workers at work. As has already been pointed out, design solutions can be generated through staff consultation because end-users interact with buildings in different ways from managers and designers.

(c) COMMISSIONING AND HAND-OVER

One of the most succinct and damning critiques of the commissioning and hand-over of UK buildings comes from Sir John Badenoch in his reports on the inquiry into the outbreak of Legionnaires' disease in Stafford hospital (Badenoch 1986, 1987). Commissioning should ensure that the whole plant performs as specified, and that operational engineers acquire the knowledge necessary to operate in the specified manner. In order to achieve this, Badenoch's observations and other literature (Lawson 1989) point to the need for:

- commissioning to be given greater importance and adequate time in the building schedule so that urgent completion dates do not prevent good practice;
- someone to take overall responsibility for commissioning;
- effective commissioning teams to be set up, capable of taking an overview of building performance rather than a narrow view of technical operation (the team should include owners/occupiers, design engineers and architects and finally, the main contractor);
- a detailed commissioning plan to be prepared (using, for example CIBSE Commissioning Codes): Lawson (1989) advises that the plan be prepared by a third party; the plan should include 'acceptance criteria';
- effective documentation to be provided, including 'as-built' drawings, operating manuals and test data;
- system familiarisation and staff training to be provided and documented.

Experience also indicates that, because buildings continue to be commissioned after they are occupied, building users are often left in the dark about changes which are being made to the system. This can lead to dissatisfaction if the system, or part of it, is being shut down for modification or is still not operating to the specification. This is clearly a difficult period for engineers, but they can retain the confidence of users if there are regular liaison meetings between system engineers and user representatives during the commissioning or 'selling-in' period. It is useful for engineers to walk though occupied areas on an occasional basis during this period to monitor occupant comfort informally and to answer any queries which may arise.

(d) OPERATION AND MAINTENANCE

In any operation, it must not be forgotten that two of the most critical operational variables are resourcing and budgeting. However, this report does not discuss financial considerations, as these are part of a more general management agenda; it focuses instead on the operational practices which can contribute to the prevention of SBS.

The demands upon operational building management are not fixed over time and therefore managers face problems which change over time: the introduction of new technology, a new way of working, a new layout arrangement or the move to a deep plan building from a narrow plan office may require the management effort to increase and alter accordingly. It has been observed, however, that managers are caught out by changes in the magnitude of problems they face. This is because the relationship between complexity and management input is not linear.

Leaman (1991) has postulated that there are thresholds of complexity in building design: if a threshold is crossed (and what constitutes a threshold will vary between buildings depending on the inherent level of adaptability), the management effort needs to be increased disproportionately. This means that buildings of high complexity may need considerably more management input than might be predicted. Leaman describes how two buildings known to be 'sick', measured as outlined by Wilson & Hedge (1987), had a management input which was far too low (relative to 'healthy' buildings of similar complexity) to cope with the building systems and services.

Managers are often unaware when they have crossed a threshold of building complexity, which may result in SBS. This points to one of the major operational instruments which building managers can use: monitoring or assembling the facts about the building. The physical variables to be monitored are indoor climate, indoor air quality and ergonomic parameters. Collecting data about these variables provides managers with a factual basis for assessing environmental quality and comparing this with the nature of staff complaints. It also provides a basis for comparing environmental quality over time.

Monitoring must be combined with an effective form of record keeping for these benefits to be obtained. The monitoring of user satisfaction, comfort and health can be done by recording and analysing complaints, or carrying out structured assessments such as an annual staff survey.

If occupant monitoring is to be carried out by means of recording complaints, then an efficient and open line of communication must be developed between building managers and users. Managers will also have to develop a response system which informs users of what action, if any, is being taken in response to their complaints. Managers should ensure that the chain of command between those who receive complaints and those who act on them is not long and that there is a means of monitoring progress. Regular analysis of complaint records is essential to reveal underlying trends which may call for a structured response.

Effective maintenance requires that appropriate intervals of maintenance are established and that there is proper documentation, which should include comprehensible, well-written instructions and manuals, detailed maintenance schedules, and efficient record keeping of the work actually carried out. Good management will result in good maintenance.

4.6.5 Conclusion

This Section has reviewed the ways in which poor management could contribute to an increased incidence of SBS. It has shown that managers can intervene to control SBS but that a significant cultural change may be required in many organisations if they are to address SBS and environmental quality as valid foci of management activity.

Within the field of management science, there is a continuing debate about the best way of achieving change in organisational culture. In a recent review of this issue, however, Pettigrew (1990) points out that. "It is rather more difficult to change the core beliefs and assumptions within the organisation than it is to change some of the manifestations of culture in, for example, the organisation's structure and systems".

With regard to SBS, different professional bodies and interested parties may wish to focus on either the willingness or the capacity of managers to act. For example, the design community may encourage organisations to place a higher value on their buildings whereas engineers may look for more technical guidance.

Practising building managers may perhaps benefit most from a reminder of the way in which one of the most respected management theorists summarises the nature of management. Table 6 has been adapted from Drucker (1955) to illustrate how efforts to control SBS map onto his view of management.

TABLE 6. THE ROLE OF MANAGEMENT (adapted from Drucker 1955)

MANAGEMENT ROLE	KEY WORDS	IMPLICATIONS FOR THE CONTROL OF SBS
Set objectives	Goals, required action, communication, balance between results & values, short & long term objectives.	Set relevant standards to control SBS (use existing best practice standards when when no others are available). Plan strategically to improve environmental design standards.
Organise	Division of tasks, allocation of responsibility & resources.	Ensure an effective chain of command in dealing with complaints; monitor their progress. Establish a maintenance schedule & keep records.
Motivate & communicate	Team building, principle of justice	Involve users in design briefing & environmental monitoring. Inform users of commissioning & maintenance work to retain their confidence in building services & the management system.
Measure	Yardsticks, performance, analysis & interpretation	Monitor environmental quality & user comfort, satisfaction & health on a regular basis. Test & commission new systems effectively.
Develop people	Training, direction self development, principle of integrity.	Give users control over their environment where possible.

5. CONCLUSION

Whilst the quality of evidence is quite variable it is useful to summarise the factors that may contribute towards symptoms associated with SBS. These are as follows.

VENTILATION RATES. In some cases symptoms have been reduced by increasing the fresh air input, although there is little evidence that this will be an effective measure where the ventilation rate already meets current guidelines. Air distribution may be a more significant problem but there is little direct evidence of this.

VENTILATION SYSTEMS. Air conditioning is strongly associated with SBS but mechanical ventilation and humidification are not. The mean figures for humidification may obscure some cases in which humidification reduces symptoms and other cases in which it increases them. The association with air conditioning is probably not due to low fresh air supply rates but may be due to poor air distribution and/or poor maintenance and/or the creation of an environment in the building which is conducive to the growth of micro-organisms and dust mites.

AIRBORNE CHEMICAL POLLUTION. Many pollutants probably contribute to SBS, the mixture depending on the age of the building and the materials used in it. Newer buildings are more likely to suffer from relatively high levels of VOCs whereas older buildings may contain the products of degradation and decay. Environmental tobacco smoke in excess would also cause symptoms.

MICRO-ORGANISMS, VIABLE AND NON-VIABLE PARTICULATES. There is increasing evidence that an important role is played by a mixture of organic and non-organic dust from ill-maintained air conditioning systems and furnishings. Mould spores or fragments, the excreta of dust mites and organic matter brought into the building on shoes and clothing may all play their part. Infection is almost certainly not involved, but allergic and irritant effects are possible.

TEMPERATURE. Temperatures above 21°C have been shown to increase symptoms, but possibly only when humidity is low or under particular conditions of air movement which either dry the eyes or provide little cooling.

HUMIDITY. Relative humidity below 30% can cause SBS symptoms but the risks due to this must be offset against the possible microbiological problems promoted by humidification.

LIGHTING. Certain symptoms may be promoted by generally poor lighting conditions, the absence of a view out of a window or by flicker from fluorescent tubes operated at 50Hz.

PERSONAL AND ORGANISATIONAL FACTORS. Symptoms are more frequent among women, workers in more routine jobs, those with a history of allergy, and those who spend many hours in VDU work. Perceived control over the indoor environment is also a significant factor.

There is little evidence of a specific role of noise or electromagnetic factors, but any source of stress or general dissatisfaction, whether from the environment, the job or the organisation, while unlikely to be a primary cause of SBS, can contribute to symptoms, or at least to the reporting of symptoms.

The preceding paragraphs summarise knowledge of the risk factors for SBS but it must be emphasised again that the evidence that these are direct causal agents is limited. It is also important to remember what may be behind these risk factors, for example:

- open plan offices of more than about 10 work stations;
- deep building plan;
- changes of use of the building and office partitioning after commissioning;
- sealed windows;
- large areas of soft furnishing and open shelving and filing;
- new furniture, carpets and painted surfaces;
- lightweight thermal properties and poor insulation;
- poor provision for daylighting and/or uncontrolled solar gain;
- no separately ventilated spaces for smoking, photocopying etc;
- services not designed for easy maintenance;
- air inlets close to exhaust or outdoor pollution source;
- inadequate filtration;
- luminaire type and position giving high glare and flicker;
- inadequate commissioning and, when necessary, recommissioning;
- poor maintenance (hygiene and operation) of building services;
- insufficient office cleaning and suspect general repair;
- poor general management and management of staff complaints;
- low general satisfaction with the job and organisation.

On the basis of current knowledge it seems unlikely that SBS symptoms will be completely eradicated in the short term since they occur to some extent even in the best current buildings. However, a certain amount can be achieved by the application of current knowledge in a more co-ordinated manner in the process of specification, design, construction, installation, commissioning and maintenance of buildings and their services. Further research will clarify the nature of the primary causes and their interactions and provide a basis for more cost-effective prevention and cure.

ACKNOWLEDGEMENTS

I am grateful for comments on earlier drafts from BRE colleagues Dr Earle Perera, John Smith, Dr Derrick Crump and Dr Les Fothergill. Thanks also to the many who have provided secretarial support, particularly Sue Winder and Catherine Horne.

REFERENCES

Abbritti, G, Accattoli, M P, Colangeli, C, Fabbri, T, Muzi, G, Fiordi, T, Dell'Omo, M and Gabrielli, A R (1990). Sick building syndrome: high prevalence in a new air conditioned building. Proceedings of the 5th International Conference on Indoor Air Quality and Climate, Toronto, Canada, 1, 513-518. Ottawa: Canada Mortgage and Housing Corporation.

Abildgaard, A (1988). The interaction between dust, microorganisms and the quality of cleaning. Berglund, B and Lindvall, T (Eds), Proceedings of Healthy Buildings 88, Stockholm, Sweden, 3, 19-23. Stockholm: Swedish Council for Building Research.

American Industrial Hygiene Conference (1987). Proceedings of the American Industrial Hygiene Conference, Montreal, Canada.

Andegiorgis, T (1988). Ventilation employing displacement flow. Berglund, B and Lindvall, T (Eds), Proceedings of Healthy Buildings 88, Stockholm, Sweden, 3, 31-37. Stockholm: Swedish Council for Building Research.

Andersen, I B, Lundqvist, G R and Proctor, D F (1973). Human perception of humidity under four controlled conditions. Archives of Environmental Health 26, 22-27.

Andersen, I, Lundqvist, G R and Mølhave, L (1985). Indoor air pollution due to chipboard used as a construction material. Atmospheric Environment 9, 1121-1127.

Anderson, I and Korsgaard, J (1984). Asthma and the indoor environment: assessment of health implications of high indoor air humidity. Berglund, B, Lindvall, T and Sundell, J (Eds), Proceedings of the 3rd International Conference on Indoor Air Quality and Climate, 1, 79-88. Stockholm: Swedish Council for Building Research.

Andersson, K, Fagerlund, I, Löfman, O and Erlandsson, B (1990). A follow-up questionnaire study after restoring modern dwellings with SBS problems. Proceedings of the 5th International Conference on Indoor Air Quality and Climate, Toronto, Canada, 1, 563-568. Ottawa: Canada Mortgage and Housing Corporation.

Andersson, L O, Frisk P, Löfstedt, B and Wyon, D P (1975). Human response to dry, humidified and intermittently humidified air. An experimental study in large office buildings. Swedish Building Research Institute Report R63:1975.

Appleby, P H and Bailey, M R (1990). Indoor air quality in Brussels meeting rooms. Proceedings of the 5th International Conference on Indoor Air Quality and Climate, Toronto, Canada, 4, 41-46. Ottawa: Canada Mortgage and Housing Corporation.

ASHRAE (1977). ASHRAE Standard 62-1973, Standards for natural and mechanical ventilation. Atlanta: American Society of Heating, Refrigerating, and Air-Conditioning Engineers, Inc.

ASHRAE (1981). ASHRAE Standard 62-1981, Ventilation for acceptable indoor air quality. Atlanta: American Society of Heating, Refrigerating, and Air-Conditioning Engineers, Inc.

ASHRAE (1989). ASHRAE Standard 62-1989, Ventilation for acceptable indoor air quality. Atlanta: American Society of Heating, Refrigerating, and Air-Conditioning Engineers, Inc.

Assael, M, Pfeifer, Y and Sulman, F G (1984). Influence of artificial air ionization on the human electroencephalogram. International Journal of Biometeorology 18, 306-312.

Axelrad, R (1989). Economic implications of indoor air quality and its regulation and control. NATO/CCMS Workshop, Pilot Study on Indoor Air Quality, The Implications of Indoor Air Quality for Modern Society, CCMS Report No. 183, 89-116. NATO/CCMS.

Badenoch, J (1986). First Report of the Committee of Inquiry into the Outbreak of Legionnaires' Disease in Stafford in April 1985. London: HMSO.

Badenoch, J (1987). Second Report of the Committee of Inquiry into the Outbreak of Legionnaires' Disease in Stafford in April 1985. London: HMSO.

Baldwin, M E and Farant, J-P (1990). Study of selected volatile organic compounds in office buildings at different stages of occupancy. Proceedings of the 5th International Conference on Indoor Air Quality and Climate, Toronto, Canada, 2, 665-670. Ottawa: Canada Mortgage and Housing Corporation.

Ball, D J (1982). A sceptic's view of air ionisation. Building Services and Environmental Engineer 5.1, 21-22.

Bayer, C W (1990). An investigation into the effect of "building bakeout" conditions on building materials and furnishings. Proceedings of the 5th International Conference on Indoor Air Quality and Climate, Toronto, Canada, 3, 581-586. Ottawa: Canada Mortgage and Housing Corporation.

Bayer, C W and Downing, C C (1991). Does a total energy recovery system provide a healthier indoor environment? Proceedings of IAQ '91 - Healthy Buildings, Washington DC, USA, 74-76. Atlanta: American Society of Heating, Refrigerating and Air-Conditioning Engineers, Inc.

Bedford, T (1974). Bedford's Basic Principles of Ventilation and Heating. London: H K Lewis and Co Ltd.

Berg, M (1989). Facial skin complaints and work at visual display units. Academic dissertation, Karolinska Institute, Stockholm. Also published in Acta Dermato-Venereologica as Supplement 150.

Berglund, B, Berglund, U, and Lindvall, T (1984). Characterisation of Indoor Air Quality and 'Sick Buildings'. Stockholm University paper AT-84-20 No 1.

Berglund, B, Berglund, U and Engen, T (1987). Do sick buildings affect Human Performance? How should one assess them? Proceedings of the 4th International Conference on Indoor Air Quality and Climate, Berlin, 2, 477-491. Berlin: Institute for Water, Soil and Air Hygiene.

Berglund, B, Johansson, I, Lindvall, T, Lundin, L and Morath, C (1988). Air quality and symptoms in a sick library. Perry, R and Kirk, P W (Eds), Proceedings of the Indoor and Ambient Air Quality Conference, 335-64. London: Selper.

Berglund, B, Johansson I, Lindvall, T and Lundin, L (1990)a. A longitudinal study of perceived air quality and comfort in a sick library building. Proceedings of the 5th International Conference on Indoor Air Quality and Climate, Toronto, Canada, 1, 489-94. Ottawa: Canada Mortgage and Housing Corporation.

Berglund, B, Johansson, I, Lindvall, T and Lundin, L (1990)b. A longitudinal study of airborne chemical compounds in a sick library building. Proceedings of the 5th International Conference on Indoor Air Quality and Climate, Toronto, Canada, 2, 677-682. Ottawa: Canada Mortgage and Housing Corporation.

Berman, S M, Greenhouse, D S, Bailey, I L, Clear, R D and Raasch, T W (1991). Human electroretinogram responses to video displays, fluorescent lighting and other high frequency sources. Optometry and Vision Science 68, 645-662.

Besch, E L and Besch, H J (1989). Indoor air quality (IAQ) problem management. ASHRAE Journal 31, 47-48.

Bischoff, E, Fischer, A, Liebenberg, B and Schirmacher, W (1990). Reduction of mites and mite excreta with agaricides - control of efficacy using new methods for excreta assessment and mite counting. Proceedings of the 5th International Conference on Indoor Air Quality and Climate, Toronto, Canada, 4, 511-516. Ottawa: Canada Mortgage and Housing Corporation.

Bischoff, E and van Bronswijk, D E M H (1986). Beiträge zur Ökologie der Hausstaubmilben I: Über die Erreichbarkeit von Hausstaubmilben durch Absaugen. Allergologie 9, 375-378.

Bishop, V L, Auster, D E and Vogel, R H (1985). The Sick Building Syndrome. What it is and how to prevent it. National Safety and Health News 132, 31-38.

Black, F W and Milroy, E A (1966). Experience of air-conditioning in offices. Journal of the Institute of Heating and Ventilating Engineers 34, 188-96.

Black, M (1990). Environmental chamber technology for the study of volatile organic compound emissions from manufactured products. Proceedings of the 5th International Conference on Indoor Air Quality and Climate, Toronto, Canada, 3, 713-718. Ottawa: Canada Mortgage and Housing Corporation.

Black, M, Pearson, W J, Brown, J, Sadie, S, Schultz, L, Peard, J, Robertson, W and Lawhon, J (1991)a. The State of Washington's experimental approach to controlling IAQ in new construction. Proceedings of IAQ '91 - Healthy Buildings, Washington DC, USA, 39-42. Atlanta: American Society of Heating, Refrigerating and Air-Conditioning Engineers, Inc.

Black, M S, Pearson, W J and Work, L M (1991)b. A methodology for determining VOC emissions from new SBR latex-backed carpet, adhesives, cushions, and installed systems and predicting their impact on indoor air quality. Proceedings of IAQ '91 - Healthy Buildings, Washington DC, USA, 267-272. Atlanta: American Society of Heating, Refrigerating and Air-Conditioning Engineers, Inc.

Bordass, W (1989). How intelligent should your office be? Paper presented to the Institute of Administrative Management Conference, 18-20 April. William Bordass Associates, London NW1.

Bordass, W (1990). Building Services. Facilities Design and Management, February, 19-25.

Bornehag, C-G (1991). Problems associated with the replacement of casein-based self-levelling compound. Proceedings of IAQ '91 - Healthy Buildings, Washington DC, USA, 273-275. Atlanta: American Society of Heating, Refrigerating and Air-Conditioning Engineers, Inc.

Boyce, P R (1981). Human Factors in Lighting. London: Applied Science Publishers.

Boyce, P R (1989). Studying sick buildings. Proceedings of Conference on Designing for Environmental Quality Conference. Birmingham Polytechnic.

Broder, I, Pilger, C and Corey, P (1990). Building related discomfort is associated with perceived rather than measured levels of indoor environmental variables. Proceedings of the 5th International Conference on Indoor Air Quality and Climate, Toronto, Canada, 1, 221-226. Ottawa: Canada Mortgage and Housing Corporation.

Broughton, A, Thrasher, J D and Madison, R (1990). Biological monitoring of indoor air pollution: a novel approach. Proceedings of the 5th International Conference on Indoor Air Quality and Climate, Toronto, Canada, 2, 145-150. Ottawa: Canada Mortgage and Housing Corporation.

Brown Skeers, V M (1984). The puzzling case of the SOB (State Office Building). Occupational Health Nursing 32, 251-254.

Brown, V M, Crump, D R, Dearling, T B, Gardiner, D and Gavin, M A (1991). The effectiveness of an air purifier for reducing concentrations of volatile organic compounds in indoor air following painting. Proceedings of IAQ '91 - Healthy Buildings, Washington DC, USA, 325-328. Atlanta: American Society of Heating, Refrigerating and Air-Conditioning Engineers, Inc.

Brundrett, G W (1975). Ventilation Requirements in Rooms Occupied by Smokers: a Review. Capenhurst, Chester: Electricity Council Research Centre.

Burge, P S (1989). Occupational risks of glutaraldehyde. British Medical Journal 299, 342.

Burge, P S (1990). Building Sickness - a medical approach to the causes. Proceedings of the 5th International Conference on Indoor Air Quality and Climate, Toronto, Canada, 5, 3-14. Ottawa: Canada Mortgage and Housing Corporation.

Burge, P S, Jones, P and Robertson, A S (1990)a. Sick building syndrome. Proceedings of the 5th International Conference on Indoor Air Quality and Climate, Toronto, Canada, 1, 479-484. Ottawa: Canada Mortgage and Housing Corporation.

Burge, P S, Robertson, A S and Hedge, A (1990)b. Validation of self-administered questionnaire in the diagnosis of sick building syndrome. Proceedings of the 5th International Conference on Indoor Air Quality and Climate, Toronto, Canada, 1, 575-580. Ottawa: Canada Mortgage and Housing Corporation.

Chandraker, K and Benhama, A (1990). Ion counter for atmospheric air. Proceedings of the 5th International Conference on Indoor Air Quality and Climate, Toronto, Canada, 3, 231-236. Ottawa: Canada Mortgage and Housing Corporation.

CIBSE (1987). SBS - Myth or Reality? Conference organised by the Chartered Institution of Building Services Engineers, Delta House, 22 Balham High Road, London SW12 9BS.

CIBSE (1991). The CIBSE Guide. London: CIBSE.

CIBSE (1984). Code for Interior Lighting. London: CIBSE.

Cole, R J and Rousseau, D (1990). Achieving environmental quality in office buildings. Proceedings of the 5th International Conference on Indoor Air Quality and Climate, Toronto, Canada, 3, 263-268. Ottawa: Canada Mortgage and Housing Corporation.

Collett, C W, Ventresca, J A and Turner, S (1991). The impact of increased ventilation on indoor air quality. Proceedings of IAQ '91 - Healthy Buildings, Washington DC, USA, 97-100. Atlanta: American Society of Heating, Refrigerating and Air-Conditioning Engineers, Inc.

Colligan, M J (1981). The psychological effect of indoor air pollution. Bulletin of the New York Academy of Medicine 57, 1014-1026, December.

Collins, B L, Fisher, W S, Gillette, G L and Marans (1989). Evaluating office lighting environments: second level analysis. USA: National Institute Standards and Technology NISTIR 89-4069.

Cooper, I (1982). Comfort and energy conservation: a need for reconciliation. Energy and Buildings 5, 83-87.

Corth, R (1983). What is "natural" light? Lighting Design and Application 13, 34-40.

COST (1989). Indoor Pollution by NO_2 in European Countries. (COST Project 613, Report 3). Commission of the European Communities Directorate General for Science, Research and Development Joint Research Centre - Institute for the Environment.

Crandall, M S, Highsmith, R, Gorman, R and Wallace, L (1990). Library of Congress and US EPA indoor air quality and work environment study: environmental survey results. Proceedings of the 5th International Conference on Indoor Air Quality and Climate, Toronto, Canada, 4, 597-602. Ottawa: Canada Mortgage and Housing Corporation.

Creuzevault, D, Cluzel, D, Dalicieux, P and Fauconnier, R (1990). An indoor air quality prediction model. Proceedings of the 5th International Conference on Indoor Air Quality and Climate, Toronto, Canada, 4, 165-170. Ottawa: Canada Mortgage and Housing Corporation.

Curwell, S and March, C (1987). Hazardous Building Materials. London: Spon.

Curwell, S, March, C and Venables, E A (1990). Health and Buildings: The Rosehaugh Guide. London: Architectural Press.

David, T A et al (1960). The sedating effect of polarised air. Proceedings of the 2nd International Bioclimatological Congress, Royal Society of Medicine, London: Pergamon Press.

Davidge, B and Kerr, G (1990). Suggestions for avoiding indoor air quality complaints. Proceedings of the 5th International Conference on Indoor Air Quality and Climate, Toronto, Canada, 3, 343-347. Ottawa: Canada Mortgage and Housing Corporation.

Davidson, J L, Black, M S, Pearson, W J, Work, L M and Miller, D P (1991). Proceedings of IAQ '91 - Healthy Buildings, Washington DC, USA, 299-303. Atlanta: American Society of Heating, Refrigerating and Air-Conditioning Engineers, Inc.

Dawson, S, Willman, P, Bamford, M and Clinton, A (1988). Safety at Work: the limits to self regulation. Cambridge University Press.

de Bortoli, M, Knöppel, H, Peil, A, Pecchio, E, Schlitt, H and de Wilde, H (1990). Investigation on the contribution of volatile organic compounds to air quality complaints in office buildings of the European Parliament. Proceedings of the 5th International Conference on Indoor Air Quality and Climate, Toronto, Canada, 2, 695-700. Ottawa: Canada Mortgage and Housing Corporation.

de Roo, F (1988). Masterplan of the interrelations between indoor environmental parameters. Berglund, B and Lindvall, T (Eds), Proceedings of Healthy Buildings 88, Stockholm, Sweden, 2, 113-119. Stockholm: Swedish Council for Building Research.

Dement, J M, Smith, N D, Hickey, J L S and Williams, T M (1984). An evaluation of formaldehyde sources, exposure and possible remedial actions in two office environments. Berglund, B, Lindvall, T, Sundell, J (Eds), Proceedings of the International conference on Indoor Air Quality and Climate, 3, 99-104. Stockholm: Swedish Council for Building Research.

Department of the Environment (1984). The Building Act 1984. London: HMSO.

Dimmick, R L and Akers, A B (1969). An Introduction to Experimental Aerobiology. New York: Wiley.

Dixon, D (1991). NORWEB Headquaters Environmental Survey. Building Research and Information 19, 147-157.

Donnini, G, Nguyen, V H and Haghighat, F (1990). Ventilation control and building dynamics by CO_2 measurement. Proceedings of the 5th International Conference on Indoor Air Quality and Climate, Toronto, Canada, 4, 257-262. Ottawa: Canada Mortgage and Housing Corporation.

Downing, C C and Bayer, C W (1991). Operation and maintenance for quality indoor air. Proceedings of IAQ '91 - Healthy Buildings, Washington DC, USA, 372-374. Atlanta: American Society of Heating, Refrigerating and Air-Conditioning Engineers, Inc.

Drake, P, Mill, P and Demeter, M (1991)a. Implications of user-based environmental control systems: three case studies. Proceedings of IAQ '91 - Healthy Buildings, Washington DC, USA, 394-400. Atlanta: American Society of Heating, Refrigerating and Air-Conditioning Engineers, Inc.

Drake, P, Mill, P, Hartkopf, V, Loftness, V, Dubin, F, Ziga, G and Posner, J (1991)b. Strategies for health promotion through user-based environmental control: a select international perspective. Proceedings of IAQ '91 - Healthy Buildings, Washington DC, USA, 14-21. Atlanta: American Society of Heating, Refrigerating and Air-Conditioning Engineers, Inc.

Dressel, D L and Francis, J (1987). Office productivity: contributions of the workstation. Behavior and information technology 6, 279-284.

Drucker, P F (1955). The Practice of Management. London: Pan.

Duffy, F (1989). The changing workplace. Architects' Journal, September 27.

Ekberg, L E (1991). Indoor air quality in a new office building. Proceedings of IAQ '91 - Healthy Buildings, Washington DC, USA, 125-127. Atlanta: American Society of Heating, Refrigerating and Air-Conditioning Engineers, Inc.

Elixmann, J H, Schata, M and Jorde, W (1990). Fungi in filters or air-conditioning-systems cause the building-related-illness. Proceedings of the 5th International Conference on Indoor Air Quality and Climate, Toronto, Canada, 1, 193-196. Ottawa: Canada Mortgage and Housing Corporation.

Environmental Protection Agency (EPA) (1971). National Primary and Secondary Air Quality Standards. Federal Register 36 (84), 88186.

Environmental Protection Agency (EPA) (1987). Indoor Air Quality Implementation Plan, Appendix A. EPA, Washington D.C.

Eto, J H (1990). The HVAC costs of increased fresh air ventilation rates in office buildings, Part 2. Proceedings of the 5th International Conference on Indoor Air Quality and Climate, Toronto, Canada, 4, 53-58. Ottawa: Canada Mortgage and Housing Corporation.

Evans, G W, and Jacobs, S (1981). Air pollution and human behaviour. Journal of Social Issues, 37, 95-125

Eysel, U T and Burandt, U (1984). Fluorescent tube light evokes flicker responses. Vision Research 24, 943-948.

Fanger, P O (1988)a. Introduction of the olf and decipol units to quantify air pollution perceived by humans indoors and outdoors. Energy and Buildings 12, 1-6.

Fanger, P O (1988)b. A comfort equation for indoor air quality and ventilation. Berglund, B and Lindvall, T (eds) Healthy Buildings 88, 1, 39-51. Stockholm: Swedish Council for Building Research.

Fanger, P O (1990). New principles for a future ventilation standard. Proceedings of the 5th International Conference on Indoor Air Quality and Climate, Toronto, Canada, July, 353-363. Ottawa: Canada Mortgage and Housing Corporation.

Fanger, P O, Melikow, A K and Hanzawa, H (1987). Draught and turbulence. Proceedings of the 4th International Conference on Indoor Air Quality and Climate, Berlin, 3, 404-408. Berlin: Institute for Water, Soil and Air Hygiene.

Fanger, P O, Lauridsen, J, Bluyssen, P and Clausen, G (1988). Air pollution sources in offices and assembly halls, quantified by the olf unit. Energy and Buildings 12, 7-19.

Farant, J-P, Bédard, S, Tamblyn, R T, Menzies, R I, Tamblyn, R M, Hanley, J and Spitzer, W O (1990). Effect of changes in the operation of a building's ventilation systems on environmental conditions at individual workstations in an office complex. Proceedings of the 5th International Conference on Indoor Air Quality and Climate, Toronto, Canada, 1, 581-585. Ottawa: Canada Mortgage and Housing Corporation.

Farant, J-P, Nguyen, V H, Leduc, J and Auger, M (1991). Impact of office design and layout on the effectiveness of ventilation provided to individual workstations in office buildings. Proceedings of IAQ '91 - Healthy Buildings, Washington DC, USA, 8-13. Atlanta: American Society of Heating, Refrigerating and Air-Conditioning Engineers, Inc.

Ferahrian, R H (1984). Indoor air pollution - some Canadian experiences. Berglund, B, Lindvall, T, Sundell, J (Eds), Proceedings of the International conference on Indoor Air Quality and Climate, 1, 207-212. Stockholm: Swedish Council for Building Research.

Fidler, A T, Wilcox, T G, Leaderer, B P, Selfridge, O J and Hornung, R W (1990). Health symptoms and comfort concerns. Proceedings of the 5th International Conference on Indoor Air Quality and Climate, Toronto, Canada, 4, 603-608. Ottawa: Canada Mortgage and Housing Corporation.

Field, A (1987). Infrasound and sick buildings. Building Services, January, 59.

Finnegan, M J, Pickering, C A C and Burge, P S (1984). The sick building syndrome: prevalence studies. British Medical Journal 289, 1573-1575.

Finnegan, M J and Pickering, C A C (1986). Building related illness. Clinical Allergy 16, 389-405.

Finnegan, M J, Pickering, A C, Gill, F S and Ashton, I (1987)a. Negative ions and the sick building syndrome. Proceedings of the 4th International Conference on Indoor Air Quality and Climate, Berlin, 2, 547-551. Berlin: Institute for Water, Soil and Air Hygiene.

Finnegan, M J, Pickering, C A C, Gill, F S, Ashton, I and Froese, D (1987)b. Effect of negative ion generators in a sick building. British Medical Journal 294, 1195-6.

Fitzner, K (1987). Air outlets for displacement flow and their influence on flow patterns with various kinds of heat sources. Proceedings of the 4th International Conference on Indoor Air Quality and Climate, Berlin, 3, 347-351. Berlin: Institute for Water, Soil and Air Hygiene.

Franck, C (1986). Eye symptoms and signs in buildings with indoor climate problems ('office eye syndrome'). Acta Opthalmologica 64, 306-11.

Franck, C and Skov, P (1990). Validation of two questionnaires used for diagnosing the sick building syndrome. Proceedings of the 5th International Conference on Indoor Air Quality and Climate, Toronto, Canada, 1, 485-488. Ottawa: Canada Mortgage and Housing Corporation.

Franzen, B (1969). Kontorsrummet - et klimat studie i nio kontorshus. Stockholm: Svensk byggtjänst.

Fredericksen, E (1990). Practical attempts to predict indoor air quality. Proceedings of the 5th International Conference on Indoor Air Quality and Climate, Toronto, Canada, 3, 239-244. Ottawa: Canada Mortgage and Housing Corporation.

Gammage, R B, Hansen, D L and Johnson, L W (1989). Indoor air quality investigations: a practitioner's approach. Environment International 15, 503-510.

Gebefuegi, I L and Korte, F (1990). Source of organics in the air of an office building. Proceedings of the 5th International Conference on Indoor Air Quality and Climate, Toronto, Canada, 2, 701-704. Ottawa: Canada Mortgage and Housing Corporation.

Girman, J R, Alevantis, L E, Petreas, M X and Webber, L M (1990). Building bake-out studies. Proceedings of the 5th International Conference on Indoor Air Quality and Climate, Toronto, Canada, 3, 349-354. Ottawa: Canada Mortgage and Housing Corporation.

Grande, L B and Hanssen, S O (1988). Allergic persons, human sensors regarding indoor environment quality? Berglund, B and Lindvall, T (Eds), Proceedings of Healthy Buildings 88, Stockholm, Sweden, 3, 561-564. Stockholm: Swedish Council for Building Research.

Graveson, S, Skov, P, Valbjørn, O and Løwenstein, H (1990). The role of potential immunogenic components of dust (MOD) in the sick-building-syndrome. Proceedings of the 5th International Conference on Indoor Air Quality and Climate, Toronto, Canada, 1, 9-13. Ottawa: Canada Mortgage and Housing Corporation.

Griffiths, W A D and Wilkinson, D S (1985). Essentials of Industrial Dermatology. Oxford: Blackwell.

Guidotti, T L, Alexander, R W and Fedoruk, M J (1987). Epidemiological features that may distinguish between building associated illness outbreaks due to chemical exposure or psychogenic origin. Journal of Occupational Medicine 29, 148-150.

Gunnarsen, L (1990). Adaptation and ventilation requirements. Proceedings of the 5th International Conference on Indoor Air Quality and Climate, Toronto, Canada, 1, 599-610, Ottawa: Canada Mortgage and Housing Corporation.

Guo, Z, Dunn, J E, Tichenor, B A, Mason, M A and Krebs, K A (1990). On representing reversible sinks in indoor air quality models. Proceedings of the 5th International Conference on Indoor Air Quality and Climate, Toronto, Canada, 4, 177-182, Ottawa: Canada Mortgage and Housing Corporation.

Gustafsson, H (1991). Building materials identified as major emission sources. Proceedings of IAQ '91 - Healthy Buildings, Washington DC, USA, 259-261. Atlanta: American Society of Heating, Refrigerating and Air-Conditioning Engineers, Inc.

Haghighat, F, Wang, J C Y and Jiang, Z (1990). Development of a three-dimensional numerical model to investigate the air flow and age distribution in a multi-zone enclosure. Proceedings of the 5th International Conference on Indoor Air Quality and Climate, Toronto, Canada, 4, 183-188. Ottawa: Canada Mortgage and Housing Corporation.

Hall, H I, Leaderer, B P, Cain, W S and Fidler, A T (1991). Influence of building-related symptoms on self-reported productivity. Proceedings of IAQ '91 - Healthy Buildings, Washington DC, USA, 33-35. Atlanta: American Society of Heating, Refrigerating and Air-Conditioning Engineers, Inc.

Hamburger, R J (1960). On the influence of artificial ionisation of the air on the oxygen uptake during exercise. Proceedings of the 2nd International Bioclimatological Congress, Royal Society of Medicine, London: Pergamon Press.

Hansen, L (1989). Monitoring of symptoms in estimating the effect of intervention in the sick building syndrome: a field study. Environment International 15, 59-62.

Hansen, T B and Andersen, B (1986). Ozone and other air pollutants from photocopying machines. American Industrial Hygiene Association Journal 47, 659-665.

Hansson, T (1988). Sick buildings - a consequence of too short a drying time? Berglund, B and Lindvall, T (Eds), Proceedings of Healthy Buildings 88, Stockholm, Sweden, 2, 305-307. Stockholm: Swedish Council for Building Research.

Harrison, J, Pickering, C A C, Faragher, E B and Austwick, P K C (1990). An investigation of the relationship between microbial and particulate indoor air pollution and the sick building syndrome. Proceedings of the 5th International Conference on Indoor Air Quality and Climate, Toronto, Canada, 1, 149-154. Ottawa: Canada Mortgage and Housing Corporation.

Hawkins, L H (1982). Air ions and office health. Occupational Health and Safety, March, 116-124.

Hawkins, L H and Barker, T (1978). Air ions and human performance. Ergonomics 21, 273-278.

Hawkins, L H and Morris, L (1984). Air ions and the sick building syndrome. Berglund, B, Lindvall, T, Sundell, J (Eds), Proceedings of the 3rd International Conference on Indoor Air Quality and Climate, 3, 197-200. Stockholm: Swedish Council for Building Research.

Hawkins, L H and Wang, T (1991). The office environment and the sick building syndrome. Proceedings of IAQ '91 - Healthy Buildings, Washington DC, USA, 365-371. Atlanta: American Society of Heating, Refrigerating and Air-conditioning Engineers, Inc.

Health and Safety Commission (1974). Health and Safety at Work Act etc. London: HMSO.

Health and Safety Executive (1983). Visual Display Units. London: HMSO.

Hedge, A (1984)a. Evidence of a relationship between office design and self-reports of ill-health among office workers in the United Kingdom. Journal of Architecture and Planning Research 3, 163-174.

Hedge, A (1984)b. Ill-health amongst office workers: an examination of the relationship between office design and employee well-being. Grandjean, E (Ed), Ergonomics and Health in Modern Offices. London: Taylor and Francis.

Hedge, A (1991). Healthy office lighting for computer workers: a comparison of lensed-indirect and direct systems. Proceedings of IAQ '91 - Healthy Buildings, Washington DC, USA, 61-66. Atlanta: American Society of Heating, Refrigerating and Air-Conditioning Engineers, Inc.

Hedge, A, Sterling, E M and Sterling, T D (1986). Evaluating Office Environments: the case for a macroergonomic systems approach. Brown, O Jr. and Hendrick, H W (Eds), Proceedings of the 2nd Symposium on Human Factors in Organisational Design and Management, 419-424, Vancouver. North Holland.

Hedge, A and Collis, M D (1987). Do negative air ions affect human mood and performance? Annals of Occupational Hygiene 31, 285-290.

Hedge, A and Ellis, D (1990). Computer-aided facilities diagnostics: a new software tool for investigating indoor environmental problems. Proceedings of the 5th International Conference on Indoor Air Quality and Climate, Toronto, Canada, 4, 127-132. Ottawa: Canada Mortgage and Housing Corporation.

Hedge, A, Erickson, W A and Rubin, G (1991)a. The effects of smoking policy on indoor air quality and sick building syndrome in 18 air-conditioned offices. Proceedings of IAQ '91 - Healthy Buildings, Washington DC, USA, 151-159. Atlanta: American Society of Heating, Refrigerating and Air-Conditioning Engineers, Inc.

Hedge, A, Martin, M G and McCarthy, J F (1991)b. Breathing-zone filtration effects on indoor air quality and sick building syndrome complaints. Proceedings of IAQ '91 - Healthy Buildings, Washington DC, USA, 351-357. Atlanta: American Society of Heating, Refrigerating and Air-Conditioning Engineers, Inc.

Hellström, B, Palmgren, U and Ström, G (1990). Four buildings with SBS-symptoms. Proceedings of the 5th International Conference on Indoor Air Quality and Climate, Toronto, Canada, 4, 385-390. Ottawa: Canada Mortgage and Housing Corporation.

Hicks, J, Worl, K and Hall, K (1990). Building bake-out during commissioning: effects on VOC concentrations. Proceedings of the 5th International Conference on Indoor Air Quality and Climate, Toronto, Canada, 3, 413-418. Ottawa: Canada Mortgage and Housing Corporation.

Hobbs, M E, Osborne, J S and Adamek, S (1956). Some compounds of gas phase of cigarette smoke. Analytical Chemistry 28, 211-215

Hodgkinson, P (1991). Mind over machine. Architects' Journal, October 30.

Hodgson, M J, Arena, V, Thorn, A, Palmer, R, Burge, H, Spengler, J, Turner, W, Fink, J N and Hemry, D (1990). Allergic tracheobronchitis in Alaska. Proceedings of the 5th International Conference on Indoor Air Quality and Climate, Toronto, Canada, 1, 197-202. Ottawa: Canada Mortgage and Housing Corporation.

Holmberg, R B, Folkesson, K, Stenberg, L-G and Jansson, G (1987). Experimental comparison of indoor climate using various air distribution methods. Proceedings of the 4th International Conference on Indoor Air Quality and Climate, Berlin, 3, 307-312. Berlin: Institute for Water, Soil and Air Hygiene.

Holt, G L (1990). Seasonal indoor/outdoor fungi ratios and indoor bacteria levels in non-compliant office buildings. Proceedings of the 5th International Conference on Indoor Air Quality and Climate, Toronto, Canada, 2, 33-38. Ottawa: Canada Mortgage and Housing Corporation.

House of Commons Environment Committee (1991). Sixth Report: Indoor Pollution. London: HMSO.

Hudnell, H K, Otto, D A, House, D E and Mølhave, L (1990). Odour and irritation effects of a volatile organic compound mixture. Proceedings of the 5th International Conference on Indoor Air Quality and Climate, Toronto, Canada, 1, 263-268. Ottawa: Canada Mortgage and Housing Corporation.

Hujanen, M, Seppänen, O and Pasanen, P (1991). Odour emission from the used filters of air-handling units. Proceedings of IAQ '91 - Healthy Buildings, Washington DC, USA, 329-333. Atlanta: American Society of Heating, Refrigerating and Air-Conditioning Engineers, Inc.

Hurrell, J J Jr, Sauter, S L, Fidler, A T, Wilcox, T G and Hornung, R W (1990). Job stress issues in the Library of Congress/EPA Headquarters indoor air quality and work environment study. Proceedings of the 5th International Conference on Indoor Air Quality and Climate, Toronto, Canada, 4, 647-652. Ottawa: Canada Mortgage and Housing Corporation.

Int Hout, L (1984). Tight building syndrome: is it hot air? Heating/Piping/Air Conditioning, January, 99-103.

International Standards Organisation (1984). ISO 7730-1984. Moderate thermal environments - Determination of the PMV and PPD indices and specification of the conditions for thermal comfort. Geneva: International Organisation for Standardisation

Ising, H, Marker, D, Shenoda, F B and Schwarze (1982). Infra-sound Effects on Humans. HdA-Reihe, VDI-Verlag Dusseldorf.

Ising, H and Schwarze, C (1982). The effects of infra-sound on humans. Zeitschrift für Larmebekampfing 29, 79-82.

Iwata, T, Doi, S and Kimura, K (1990). Effect of adhered tobacco smoke on odor sensation. Proceedings of the 5th International Conference on Indoor Air Quality and Climate, Toronto, Canada, 1, 349-354. Ottawa: Canada Mortgage and Housing Corporation.

Jaakkola, J J K (1986). Toimistorakennuksen sisäilma ja terveys. Kokeellinen ja epidemiologinen tutkimus koneellisen ilmanvaihdon vaikutuksista. Väitöskirja. (English summary: Indoor air in office building and human health. Experimental and epidemiologic study of the effects of mechanical ventilation.) Doctoral thesis. Department of Public Health, University of Helsinki. Helsinki: Health Services Research by the National Board of Health in Finland.

Jaakkola, J J K, Heinonen, O P and Seppanen, O (1989). Sick building syndrome, sensation of dryness and thermal comfort in relation to room temperature in an office building: need for individual control of temperature. Environment International 15, 163-16.

Jaakkola, J J K, Miettinen, O S, Komulainen, K, Tuomaala, P and Seppänen (1990)a. The effect of air recirculation on symptoms and environmental complaints in office workers. A double-blind, four period cross-over study. Proceedings of the 5th International Conference on Indoor Air Quality and Climate, Toronto, Canada, 1, 281-286. Ottawa: Canada Mortgage and Housing Corporation.

Jaakkola, J J K, Reinikainen, L M, Heinonen, O P, Majanen, A and Seppänen, O (1990)b. Indoor air quality requirements for healthy office buildings: recommendations based on an epidemiologic study. Paper presented at Conseil Internationale du Bâtiment W77 meeting, Rotterdam.

Jackman, P J (1991). Displacement ventilation. Proceedings of the CIBSE National Conference, University of Kent, Canterbury, 364-371. London: Chartered Institution of Building Services Engineers.

Janssen, J E, and Hill, T J (1982). Ventilation for control of indoor air quality: a case study. Environment International 8, 487-496.

Jantunen, M J, Bunn, E, Pasanen, P and Pasanen, A-L (1990). Does moisture condensation in air ducts promote fungal growth? Proceedings of the 5th International Conference on Indoor Air Quality and Climate, Toronto, Canada, 2, 73-78. Ottawa: Canada Mortgage and Housing Corporation.

Johansson, G and Aronsson, G (1984). Stress reactions in computerised administrative work. Journal of Occupational Behaviour 5, 159-181.

Johnsen, C R, Heinig, J H, Schmidt, K, Albrechtsen, O, Nielsen, P A , Nielsen, G D, Hansen, L F, Wolkoff, P and Frank, C (1990). Proceedings of the 5th International Conference on Indoor Air Quality and Climate, Toronto, Canada, 1, 269-274. Ottawa: Canada Mortgage and Housing Corporation.

Johnson, A R (1990). Formaldehyde sensitivity. Proceedings of the 5th International Conference on Indoor Air Quality and Climate, Toronto, Canada, 1, 43-48. Ottawa: Canada Mortgage and Housing Corporation.

Kerr, G and Sauer, P (1990). Control strategies for liquid process photocopier emissions. Proceedings of the Fith International Conference on Indoor Air Quality and Climate, Toronto, Canada, 3, 759-770. Ottawa: Canada Mortgage and Housing Corporation.

Kim, S D, Yamamoto, T, Ensor, D S and Sparks L E (1990). Three-dimensional containment distribution in an office space. Proceedings of the 5th International Conference on Indoor Air Quality and Climate, Toronto, Canada, 4, 139-144. Ottawa: Canada Mortgage and Housing Corporation.

Kirkbride, J (1985). Sick Building Syndrome: Causes and Effects. Presented to Human Resources Division, Ottawa Civic Hospital, Ottawa, Ontario, October.

Kjær, U and Nielsen, P A (1991). Adsorption and desorption of organic compounds on fleecy materials. Proceedings of IAQ '91 - Healthy Buildings, Washington DC, USA, 285-288. Atlanta: American Society of Heating, Refrigerating and Air-Conditioning Engineers, Inc.

Kjaergaard, S, Mølhave, L and Pedersen (1990)a. Changes in human sensory reactions, eye physiology, and performance when exposed to a mixture of 22 indoor air volatile organic compounds. Proceedings of the 5th International Conference on Indoor Air Quality and Climate, Toronto, Canada, 1, 319-324. Ottawa: Canada Mortgage and Housing Corporation.

Kjaergaard, S, Pederson, O F and Mølhave, L (1990)b. Common chemical sense of the eyes - influence of smoking, age, and sex. Proceedings of the 5th International Conference on Indoor Air Quality and Climate, Toronto, Canada, 1, 257-262. Ottawa: Canada Mortgage and Housing Corporation.

Kröling, P (1983). Disturbances to well-being and comfort in air-conditioned buildings: studies of the "building illness syndrome". Zuckschwerdt-Verlag Munich, Berne, Vienna (Identical with Final Report for the Federal Ministry for Research and Technology, KFZ Nr.01 VD 1132).

Kröling, P (1987). Untersuchungen zum 'Building Illness'-Syndrom in klimatisierten Gebäuden. Gesundheits Ingenieur Haustechnik Bauphysik Umwelttechnik 108.

Kröling, P (1988). Health and well-being disorders in air-conditioned buildings; comparative investigations of the "building illness syndrome". Energy and Buildings 11, 277-282.

Kroner, W M (1988). A New Frontier: Environments for Innovation. Proceedings of the International Symposium on Advanced Comfort Systems for the Work Environment, May.

Krueger, A P and Reed, E J (1976). Biological impact of small air ions. Science 193, 1209-1213.

Krueger, A P and Smith, R F (1960). The biological effects of air ion action, II: negative air ion effects on the concentration and metabolism of 5-hydroxytriptamine in the mammalian respiratory tract. Journal of General Physiology 44, 269-276.

Landrus, G and Witherspoon, J (1990). Do windows make a difference? A longitudinal study of effects of installing openable windows. Proceedings of the 5th International Conference on Indoor Air Quality and Climate, Toronto, Canada, 1, 611-616. Ottawa: Canada Mortgage and Housing Corporation.

Last, J M (1983). A Dictionary of Epidemiology. Oxford: University Press.

Lawson, C N (1989). Commissioning and Indoor Air Quality. ASHRAE Journal 31, 56-63.

Leaderer, B, Wilcox, T, Fidler, A, Selfridge, J, Hurrel, J, Kollanander, M, Clickner, R, Fine, L and Teichman, K (1990). Protocol for a comprehensive investigation of building related complaints. Proceedings of the 5th International Conference on Indoor Air Quality and Climate, Toronto, Canada, 4, 609-614. Ottawa: Canada Mortgage and Housing Corporation.

Leaman, A (1991). What is a building for? Paper presented to the Building Pathology 91 Conference at Trinity College Oxford. Available from Building Use Studies, London.

Leinster, P (1990). Final Contract Report to BRE: Pilot Study of SBS.

Leinster, P, Raw, G, Thomson, N, Leaman, A and Whitehead, C (1990). A modular longitudinal approach to the investigation of sick building syndrome. Proceedings of the 5th International Conference on Indoor Air Quality and Climate, Toronto, Canada, 1, 287-292. Ottawa: Canada Mortgage and Housing Corporation.

Lepage, M F and Schuyler, G D (1990). How fresh is fresh air? Proceedings of the 5th International Conference on Indoor Air Quality and Climate, Toronto, Canada, 4, 311-316. Ottawa: Canada Mortgage and Housing Corporation.

Levin, H (1990). Critical building design factors for indoor air quality and climate: current status and predicted trends. Proceedings of the 5th International Conference on Indoor Air Quality and Climate, Toronto, Canada, 3, 301-306. Ottawa: Canada Mortgage and Housing Corporation.

Lindén, V and Rolfsen, S (1981). Video computer terminals and occupational dermatitis. Scandinavian Journal of Work and Environmental Health 7, 62-67.

Lindvall, T (1985)a. Exposure limits for office environments. International Symposium on Occupational Exposure Limits, Copenhagen, April.

Lindvall, T I (1985)b. Exposure limits for office environments. Annals of the American Conference of Industrial Hygiene 12, 99-108.

McDonald, J C, Arhirii, M, Armstrong, B, Benard, J, Cherry, N M, Cyr, D, Farant, J P, McKenna, T and McKinnon, D (1986). Building illness in a large office complex. Walkinshaw, D S (Ed), Indoor Air Quality in Cold Climates: Hazards and Abatement Measures, 7-22. Ottawa: APCA.

MacDonald, P (1991). Impact of air filters, heat exchangers, humidifiers and mixing sections on indoor air quality. Roulet, C A (Ed), Proceedings of a Workshop on Indoor Air Quality Management, Lausanne, 151-159. Luxembourg: Commission of the European Communities, Scientific and Technical Communication Unit.

McIntyre, D A (1975). Subjective Responses to Relative Humidity: A Second Experiment. Capenhurst, Chester: The Electricity Council Research Centre.

McLaughlin, P and Aigner, R (1990). Higher alcohols as indoor air pollutants: source, cause, mitigation. Proceedings of the 5th International Conference on Indoor Air Quality and Climate, Toronto, Canada, 3, 587-591. Ottawa: Canada Mortgage and Housing Corporation.

Magoun H W (1958). The Waking Brain. Springfield, Illinois: Charles C Thomas.

Mainville, C, Auger, P L, Smoragiewicz, W, Neculcea, D, Neculcea, J and Lévesque, M (1988). Mycotoxines et syndrome d'extreme fatigue dans un hopital. Berglund, B and Lindvall, T (Eds), Proceedings of Healthy Buildings 88, Stockholm, Sweden, 2, 309-317. Stockholm: Swedish Council for Building Research.

Markus, T A (1967). The significance of sunshine and view for office workers. Hopkinson, R G (Ed), Sunshine Buildings. Rotterdam: Bouwcentrum International.

Martin, P J (1991). Criteria for selection of soft furnishings for healthy buildings. Proceedings of IAQ '91 - Healthy Buildings, Washington DC, USA, 210-212. Atlanta: American Society of Heating, Refrigerating and Air-Conditioning Engineers, Inc.

Martikainen, P J, Asikainen, A, Nevalainen, A, Jantunen, M, Pasanen, P and Kalliokoski, P (1990). Proceedings of the 5th International Conference on Indoor Air Quality and Climate, Toronto, Canada, 3, 203-206. Ottawa: Canada Mortgage and Housing Corporation.

Matsunaga, K, Hayakawa, R, Ono, Y and Hisanaga, O (1988). Facial rash in visual display terminal operator. Annual Report of the Nagoya British Hospital 22, 57-61.

Matthews, R (1985). Pollution begins at home. New Scientist 108, 34-37.

Matthews, T G, Reed, T J, Tromberg, B J, Daffron, C R and Hawthorne, A R (1984). Formaldehyde emissions from consumer and construction products: potential impact on indoor formaldehyde concentrations. Berglund, B, Lindvall, T, Sundell, J (Eds), Proceedings of the 3rd International Conference on Indoor Air Quality and Climate, 3, 115-120. Stockholm: Swedish Council for Building Research.

Mayer, E and Schwab, R (1988). Direction of low turbulent air flow and thermal comfort. Berglund, B and Lindvall, T (Eds), Proceedings of Healthy Buildings 88, Stockholm, Sweden, 2, 577-582. Stockholm: Swedish Council for Building Research.

Mendell, M J (1990). Elevated symptom prevalence in air-conditioned office buildings: a reanalysis of epidemic studies from the United Kingdom. Proceedings of the 5th International Conference on Indoor Air Quality and Climate, Toronto, Canada, 1, 623-628. Ottawa: Canada Mortgage and Housing Corporation.

Menzies, R, Tamblyn, R, Farant, J P, Hanley, J, Tamblyn, R T, Nunes, F and Marcotte, P (1991). The effect of varying levels of outdoor ventilation on symptoms of sick building syndrome. Proceedings of IAQ '91 - Healthy Buildings, Washington DC, USA, 90-96. Atlanta: American Society of Heating, Refrigerating and Air-Conditioning Engineers, Inc.

Meyer, B (1984). Formaldehyde release from building products. Berglund, B, Lindvall, T and Sundell, J (Eds), Proceedings of the 3rd International Conference on Indoor Air Quality and Climate, 3, 29-34. Stockholm: Swedish Council for Building Research.

Mølhave, L (1985). Volatile organic compounds as indoor air pollutants. Gammage, R B and Kaye, S V (Eds), Indoor Air and Human Health. Chelsea, Michigan: Lewis Publishers.

Mølhave, L (1990). Volatile organic compounds, indoor air quality and health. Proceedings of the 5th International Conference on Indoor Air Quality and Climate, Toronto, Canada, 5, 15-33. Ottawa: Canada Mortgage and Housing Corporation.

Molina, C, Pickering, C A C, Valbjørn, O and de Bortoli, M (1989). Sick building syndrome - a practical guide. Commission of the European Communities COST 613 Report No.4, ECSC-EEC-EAEC, Brussels/Luxembourg.

Morey, P and MacPhaul, D (1990). Rank order assessment of volatile organic compounds in indoor air quality evaluations. Proceedings of the 5th International Conference on Indoor Air Quality and Climate, Toronto, Canada, 2, 735-739. Ottawa: Canada Mortgage and Housing Corporation.

Morey, P and Williams, C (1990). Porous insulation in buildings: a potential source of microorganisms. Proceedings of the 5th International Conference on Indoor Air Quality and Climate, Toronto, Canada, 4, 529-533. Ottawa: Canada Mortgage and Housing Corporation.

Morey, P R and Williams, C M (1991). Is porous insulation inside an HVAC system compatible with a healthy building? Proceedings of IAQ '91 - Healthy Buildings, Washington DC, USA, 128-135. Atlanta: American Society of Heating, Refrigerating and Air-Conditioning Engineers, Inc.

Nagda, N L, Koontz, M D and Albrecht, R J (1991). Effect of ventilation rate in a healthy building. Proceedings of IAQ '91 - Healthy Buildings, Washington DC, USA, 101-107. Atlanta: American Society of Heating, Refrigerating and Air-Conditioning Engineers, Inc.

Naismith, O F (1991). Electromagnetic Fields: a Review of the Evidence for Effects on Health. BRE Report 206. Garston: BRE.

Nelson, C J, Leaderer, B P, Teichman, K, Wallace, L, Kollander, M and Clickner, R P (1990). Environmental Protection Agency indoor air quality and work environment study: health symptoms and comfort concerns. Proceedings of the 5th International Conference on Indoor Air Quality and Climate, Toronto, Canada, 4, 615-620. Ottawa: Canada Mortgage and Housing Corporation.

Nelson, C J, Clayton, C A, Wallace, L A, Highsmith, V R, Kollander, M, Bascom, R and Leaderer, B P (1991). EPA's indoor air quality and work environment survey: relationships of employees' self-reported health symptoms with direct indoor air quality measurements. Proceedings of IAQ '91 - Healthy Buildings, Washington DC, USA, 22-32. Atlanta: American Society of Heating, Refrigerating and Air-Conditioning Engineers, Inc.

Nevalainen, A, Kotimaa, M, Pasanen, A L, Pellikka, M, Niininen, M, Reponen, T and Kalliokoski, P (1990). Mesophilic actinomycetes - the real indoor air problem? Proceedings of the 5th International Conference on Indoor Air Quality and Climate, Toronto, Canada, 1, 203-206. Ottawa: Canada Mortgage and Housing Corporation.

Nielsen, P A (1988). The importance of building materials and building construction to the 'sick buiilding syndrome'. Berglund, B and Lindvall, T (Eds), Proceedings of Healthy Buildings 88, Stockholm, Sweden, 3, 391-399. Stockholm: Swedish Council for Building Research.

NIOSH (1987). Guidance for Indoor Air Quality Investigations. Cincinnati: NIOSH.

Norbäck, D and Edling, C (1991). Environmental, occupational, and personal factors related to the prevalence of sick building syndrome in the general population. British Journal of Industrial Medicine 48, 451-462.

Norbäck, D and Torgen, M (1987). A longitudinal study of symptoms associated with wall-to-wall carpets and electrostatic charge in swedish school buildings. Proceedings of the 4th International Conference on Indoor Air Quality and Climate, Berlin, 2, 572-576. Berlin: Institute for Water, Soil and Air Hygiene.

Norbäck D and Torgen, M (1989). A longitudinal study relating carpeting with sick building syndrome. Environment International 15, 129-36.

Nordic Committee on Building Regulations (NKB) (1981). Indoor climate, NKB Report No 41, 76-106. Civiltryck AB Stockholm.

Nordic Committee on Building Regulations (NKB) (1991). Indoor climate - air quality. Report No 61E. Helsinki, ISBN 951-47-5322-4.

Occupational Safety and Health Administration (OSHA) (1975). Code of Federal Regulations, 40, 23072. US Department of Labor.

Otto, D, Mølhave, L, Rose, G, Hudnell, H K and House, D (1990). Neurobehavioral and sensory irritant effects of controlled exposure to a complex mixture of volatile organic compounds. Neurotoxicology and Teratology 12, 649-652.

Owen, M K, Ensor, D S and Sparks, L E (1990). Airborne particle sizes and sources found in indoor air. Proceedings of the 5th International Conference on Indoor Air Quality and Climate, Toronto, Canada, 2, 79-84. Ottawa: Canada Mortgage and Housing Corporation.

Palonen, J, Majanen, A and Seppanen, O (1988). Performance of displacement air distribution system in a small office room. Berglund, B and Lindvall, T (Eds), Proceedings of Healthy Buildings 88, Stockholm, Sweden, 3, 103-112. Stockholm: Swedish Council for Building Research.

Palonen, J and Seppänen, O (1990). Design criteria for central ventilation and air-conditioning system of offices in cold climate. Proceedings of the 5th International Conference on Indoor Air Quality and Climate, Toronto, Canada, 4, 299-304. Ottawa: Canada Mortgage and Housing Corporation.

Pasanen, P, Tarhanen, J, Kalliokoski, P and Nevalainen, A (1990). Emissions of volatile organic compounds from air conditioning filters of office buildings. Proceedings of the 5th International Conference on Indoor Air Quality and Climate, Toronto, Canada, 3, 183-186. Ottawa: Canada Mortgage and Housing Corporation.

Pejtersen, J, Bluyssen, P M, Kondo, H, Clausen, G and Fanger, P O (1989). Air pollution sources in ventilation systems. CLIMA 2000, Sarajevo, 3, 139-144.

Perera, M D A E S, Tull, R G and Walker, R R (1988). Natural ventilation for a crown court: developing statistical assessment techniques at the design stage. Proceedings of the 9th AIVC Conference on Effective Ventilation, Gent, Belgium.

Persily, A and Dols, W S (1990). Ventilation and air quality investigation of the Madison Building: preliminary results. Proceedings of the 5th International Conference on Indoor Air Quality and Climate, Toronto, Canada, 4, 621-626. Ottawa: Canada Mortgage and Housing Corporation.

Pettigrew, A (1990). Is corporate culture manageable? Wilson, D C and Rosenfeld, R H (Eds), Managing Organisations. McGraw Hill: Maidenhead.

Pickering, C A C, Finnegan, M J, Robertson, A and Burge, S (1984). Sick building syndrome. Berglund, B, Lindvall, T, Sundell, J (Eds), Proceedings of the 3rd International Conference on Indoor Air Quality and Climate, 3, 321-325. Stockholm: Swedish Council for Building Research.

Pimm, P E, Shephard, R J and Silverman, F (1978). Physiological effects of acute passive exposure to cigarette smoke. Archives of Environmental Health 33, 201-214.

Preller, L, Zweers, T, Brunekreef and Boleij, J S M (1990). Sick leave due to work-related health complaints among office workers in the Netherlands. Proceedings of the 5th International Conference on Indoor Air Quality and Climate, Toronto, Canada, 1, 227-230. Ottawa: Canada Mortgage and Housing Corporation.

Prezant, B, Bearg, D and Turner, W (1990). Tailoring lease specifications, proposals and work letter agreements to maximize indoor air quality. Proceedings of the 5th International Conference on Indoor Air Quality and Climate, Toronto, Canada, 3, 371-376. Ottawa: Canada Mortgage and Housing Corporation.

Public Works Canada (1990). Managing Indoor Air Quality. Ottawa: Public works Canada.

Putnam, V L, Woods, J E and Rask, D R (1990). A comparison of carbon dioxide concentrations and indoor environmental acceptability in commercial buildings. Proceedings of the 5th International Conference on Indoor Air Quality and Climate, Toronto, Canada, 3, 365-370. Ottawa: Canada Mortgage and Housing Corporation.

Raatschen, W (1991). IAQ-management by demand controlled ventilation. Roulet, C A (Ed), Proceedings of a Workshop on Indoor Air Quality Management, 181-190, Lausanne. Luxembourg: Commission of the European Communities, Scientific and Technical Communication Unit.

Rajhans, G S (1983). Indoor Air Quality and CO_2 Levels. Occupational Health in Ontario, 4, No 4. Ottawa: Ontario Ministry of Labour.

Rask, D R and Lane, C A (1989). Correcting maintenance deficiencies to resolve sick building syndrome. ASHRAE Journal 5, 54-56.

Rasmussen, O B (1971). Man's subjective perception of air humidity. 5th International Conference on Heating, Ventilation and Air Conditioning, 79-86.

Raw, G J, Roys, M S and Leaman, A (1990). Further findings from the office environment survey: productivity. Proceedings of the 5th International Conference on Indoor Air Quality and Climate, Toronto, Canada, 1, 231-236. Ottawa: Canada Mortgage and Housing Corporation.

Raw, G J, Leinster, P, Thompson, N, Leaman A and Whitehead, C (1991). A new approach to the investigation of sick building syndrome. Proceedings of the CIBSE National Conference, 339-343. London: The Chartered Institution of Building Services Engineers.

Raza, S H, Shylaja, G, Murthy, M S R and Bhagya Lakshmi, O (1990). Response of succulents to indoor CO_2 concentration in different habitations. Proceedings of the 5th International Conference on Indoor Air Quality and Climate, Toronto, Canada, 3, 219-224. Ottawa: Canada Mortgage and Housing Corporation.

Reinikainen, L M, Jaakkola, J J K and Heinonen, O P (1988). The effect of air humidification on different symptoms in an office building. An epidemiological study. Berglund, B and Lindvall, T (Eds), Proceedings of Healthy Buildings 88, Stockholm, Sweden, 3, 207-215. Stockholm: Swedish Council for Building Research.

Reinikainen, L M, Jaakkola, J J K, Helenius, T and Seppänen, O (1990). The effect of air humidification on symptoms and environmental complaints in office workers. A six period cross-over study. Proceedings of the 5th International Conference on Indoor Air Quality and Climate, Toronto, Canada, 1, 775-780. Ottawa: Canada Mortgage and Housing Corporation.

Reisenberg, D E and Arehard-Treichel, J (1986). Sick building syndrome plagues workers, dwellers. Journal of the American Medical Association 255, 3063.

Rhodes, W W and Gilyard, Y O (1990). Efficacy studies of an antimicrobial coating agent used in a building interior and the air handling systems to lower microbial contamination. Proceedings of the 5th International Conference on Indoor Air Quality and Climate, Toronto, Canada, 4, 559-564. Ottawa: Canada Mortgage and Housing Corporation.

Rim, Y (1977). Psychological test performance of different personality types on sharov days in artificial air ionisation. International Journal of Biometeorology 21, 337-320.

Robertson, A S and Burge, P S (1985). Building Sickness. The Practitioner 229, 531-534.

Robertson, A and Burge, S (1986). Building sickness - all in the mind? Occupational Health, March, 78-81.

Robertson, A S, Burge, P S, Hedge, A, Sims, J, Gill, F S, Finnegan, M, Pickering, C A C and Dalton, G (1985). Comparison of health problems related to work and environmental measurement in two office buildings with different ventilation systems. British Medical Journal 291, 373-376.

Robertson, A S, Burge, P S, Hedge, A, Wilson, S and Harris, J (1988). The relationship between passive cigarette smoke exposure in office workers and symptoms of "building sickness". Proceedings of Conference of Indoor and Ambient Air Quality, 320-326, Imperial College London.

Robertson, A S, McInnes, M, Glass, D, Dalton, G and Burge, P S (1989). Building sickness, are symptoms related to the office lighting? Annals of occupational Hygiene 33, 47-59.

Robertson, A S, Roberts, K T, Burge, P S and Raw, G J (1990). The effects of change in building ventilation category on sickness absence rates and the prevalence of sick building syndrome. Proceedings of the 5th International Conference on Indoor Air Quality and Climate, Toronto, Canada, 1, 237-242. Ottawa: Canada Mortgage and Housing Corporation.

Robertson, G (1988). Source, nature and symptomatology of indoor air pollutants. Berglund, B. and Lindvall, T. (Eds): Healthy Buildings 88, 3, 507-516. Stockholm: Swedish Council for Building Research.

Robertson, K A , Ghosh, T K, Hines, A L, Loyalka, S K, Warder, R C Jr and Novosel, D (1990)b. Airborne microorganisms: their occurrence and removal. Proceedings of the 5th International Conference on Indoor Air Quality and Climate, Toronto, Canada, 4, 565-570. Ottawa: Canada Mortgage and Housing Corporation.

Rodberg, J A, Miller, J F, Keller, G E and Woods, J E (1991). A novel technique to permanently remove indoor air pollutants. Proceedings of IAQ '91 - Healthy Buildings, Washington DC, USA, 311-317. Atlanta: American Society of Heating, Refrigerating and Air-Conditioning Engineers, Inc.

Rosell, L (1990). High levels of a semi-VOC in indoor air due to emission from vinyl floorings. Proceedings of the 5th International Conference on Indoor Air Quality and Climate, Toronto, Canada, 3, 707-712. Ottawa: Canada Mortgage and Housing Corporation.

Ruys, T (1970). Windowless offices. MA Thesis, University of Washington, Seattle, Washington, USA.

Rycroft, R J G and Calnan, C D (1984). Facial rashes among visual display unit operators. Pearce, B G (Ed), Health hazards of VDTs? 13-15. Chichester: Wiley and Sons.

Saarela, K and Sandell, E (1991). Comparative emission studies of flooring materials with reference to Nordic guidelines. Proceedings of IAQ '91 - Healthy Buildings, Washington DC, USA, 262-265. Atlanta: American Society of Heating, Refrigerating and Air-Conditioning Engineers, Inc.

Sandström, M, Mild, K H, Lönnberg, G, Stenberg, B, Sundell, J, Zingmark, P A and Wall, S (1990). The office illness project in Northern Sweden: IV skin symptoms among VDT workers related to electromagnetic fields - a case referent study. Proceedings of the 5th International Conference on Indoor Air Quality and Climate, Toronto, Canada, 4, 659-663. Ottawa: Canada Mortgage and Housing Corporation.

Schaeffler, A, Schultz, U and Beckert, J (1988). Carry over of pollutants in rotary air-to-air heat exchangers (regenerative heat recovery systems). Berglund, B and Lindvall, T (Eds), Proceedings of Healthy Buildings 88, Stockholm, Sweden, 3, 113-119. Stockholm: Swedish Council for Building Research.

Schiefer, H B (1990). Mycotoxins in indoor air: a critical toxicological viewpoint. Proceedings of the 5th International Conference on Indoor Air Quality and Climate, Toronto, Canada, 1, 167-172. Ottawa: Canada Mortgage and Housing Corporation.

Schriever, E and Marutzky, R (1990). VOC emissions of coated parqueted floors. Proceedings of the 5th International Conference on Indoor Air Quality and Climate, Toronto, Canada, 3, 551-555. Ottawa: Canada Mortgage and Housing Corporation.

Schulte-Hillen, G (1991). Quoted in the Architects' Journal, June 26.

Schultz, D (1965). Sensory Restriction, Effects on Behaviour. Academic Press.

Seifert, B (1990). Regulating indoor air. Proceedings of the 5th International Conference on Indoor Air Quality and Climate, Toronto, Canada, 5, 35-49. Ottawa: Canada Mortgage and Housing Corporation.

Selfridge, O J, Berglund, L G and Leaderer, B P (1990). Thermal comfort dissatisfaction responses in the Library of Congress and Environmental Protection Agency indoor air quality and work environment study. Proceedings of the 5th International Conference on Indoor Air Quality and Climate, Toronto, Canada, 4, 665-670. Ottawa: Canada Mortgage and Housing Corporation.

Shaughnessy, R J and Oatman, L (1991). The use of ozone generators for the control of indoor air contaminants in an occupied environment. Proceedings of IAQ '91 - Healthy Buildings, Washington DC, USA, 318-24. Atlanta: American Society of Heating, Refrigerating and Air-Conditioning Engineers, Inc.

Shen, J (1990). Fresh air is provided into air conditioned room directly. Proceedings of the 5th International Conference on Indoor Air Quality and Climate, Toronto, Canada, 3, 199-201. Ottawa: Canada Mortgage and Housing Corporation.

Shephard, R J, Collins, R and Silverman, F (1979)a. Responses of exercising subjects to acute "passive" cigarette smoke exposure. Environment Research 19, 279-291.

Shephard, R J, Collins, R and Silverman, F (1979)b. "Passive" exposure of asthmatic subjects to cigarette smoke. Environment Research 20, 392-402.

Shumate, M W and Wilhelm, J E (1990). Fiber shedding characteristics of commercial air filtration media. Proceedings of the 5th International Conference on Indoor Air Quality and Climate, Toronto, Canada, 3, 157-162. Ottawa: Canada Mortgage and Housing Corporation.

Skov, P and Valbjørn, O (1987)a. The "sick" building syndrome in the office environment. Proceedings of the 4th International Conference on Indoor Air Quality and Climate, Berlin, 1, 439-443. Berlin: Institute for Water, Soil and Air Hygiene.

Skov P and Valbjørn O (1987)b. "Sick" building syndrome in the office environment: The Danish Town Hall Study. Environment International 13, 339-49.

Skov, P and Valbjørn, O (1990). The Danish Town Hall study - a one-year follow-up. Proceedings of the 5th International Conference on Indoor Air Quality and Climate, Toronto, Canada, 1, 787-791. Ottawa: Canada Mortgage and Housing Corporation.

Skov, P, Valbjørn, O and Danish Indoor Climate Study Group (1987). The "sick" building syndrome in the office environment: the Danish Town Hall study. Environment International 13, 339-349.

Skov, P, Valbjørn, O, Pedersen, B V and the Danish Indoor Climate Study Group (1989). Influence of personal characteristics, job-related factors and psychosocial factors on the sick building syndrome. Scandinavian Journal of Work and Environmental Health 15, 286-295.

Smith, D D, Donovan, R P, Ensor, D S and Sparks, L E (1990). Quantification of particulate emission rates from vacuum cleaners. Proceedings of the 5th International Conference on Indoor Air Quality and Climate, Toronto, Canada, 3, 647-652. Ottawa: Canada Mortgage and Housing Corporation.

Smith, J and Webb, B (1991). Humidification: friend or foe? Building Services 13, 45.

Smith, M J, Colligan, M J and Hurell, J J (1978). Three incidents of industrial mass psychogenic illness. Journal of Occupational Medicine 20, 399-401.

Sparks, L, et al (1989). Verification and uses of the Environmental Protection Agency's indoor air model. The human equation - health and comfort. Proceedings of IAQ '89, San Diego, California, 146-150. Atlanta: the American Society of Heating, Refrigerating, and Air-Conditioning Engineers, Inc.

Sparks, L E (1991). An analysis of IAQ control options and the effects of sources and sinks. Proceedings of IAQ '91 - Healthy Buildings, Washington DC, USA, 289-291. Atlanta: American Society of Heating, Refrigerating and Air-Conditioning Engineers, Inc.

Sparks, L E and Tucker, W G (1990). A computer model for calculating individual exposure due to indoor air pollution sources. Proceedings of the 5th International Conference on Indoor Air Quality and Climate, Toronto, Canada, 4, 213-218. Ottawa: Canada Mortgage and Housing Corporation.

Stanwell-Smith, R (1986). The sick building syndrome; discussion of the problem and an account of an outbreak. Seminar on building-related illnesses. London: Crown Eagle Communications Ltd.

Steele, F (1986). Making and Managing High-Quality Workplaces. New York: Teachers College Press.

Stellman, M J, et al (1985). Air quality and ergonomics in the office: survey results and metholologic issues. American Industrial Hygiene Association Journal 46, 286-293.

Stenberg, B (1987). A rosacea-like skin rash in VDU-operators (Knave, B, Widdebäck, P-D, eds). Work with display units, 160-164. Amsterdam: Elsevier Science Publishers BV.

Stenberg, B, Lönnberg, G and Wall, S (1990)a. The office illness project in Northern Sweden. Part II: case referent study of sick building syndrome (SBS) and VDT related skin symptoms - clinical characteristics of cases and pre-disposing factors. Proceedings of the 5th International Conference on Indoor Air Quality and Climate, Toronto, Canada, 4, 683-688. Ottawa: Canada Mortgage and Housing Corporation.

Stenberg, B, Mild, K H, Sandström, M, Lönnberg, G, Wall, S, Sundell, J and Zingmark, P A (1990)b. The Office Illness project in Northern Sweden. Part I: a prevalence study of sick building syndrome (SBS) related to demographic data, work characteristics and building factors. Proceedings of the 5th International Conference on Indoor Air Quality and Climate, Toronto, Canada, 4, 627-632. Ottawa: Canada Mortgage and Housing Corporation.

Sterling, E and Sterling, T (1983)a. The impact of different ventilation levels and fluorescent lighting types on building illness: an experimental study. Canadian Journal of Public Health 74, 385-92.

Sterling, E M and Sterling, T D (1983)b. Health and comfort problems in air-conditioned buildings. Second International Congress on Building Energy Management, Iowa, USA, 7B.1-7B.10.

Sterling, E M, McIntyre, E D, Collett, C W, Sterling, T D and Meredith, J (1985). Sick buildings: case studies of tight building syndrome and indoor air quality investigations in modern office buildings. Environmental Health Review 29, 11-19.

Sterling, T D, Collett, C W and Sterling, E M (1987). Environmental tobacco smoke and indoor air quality in modern office work environments. Journal of Occupational Medicine 29, 57-62.

Stockton, M B, Spaite, P S, McLean, J S, White, J B and Jackson, M D (1991). Catalogue of materials as potential sources of indoor air pollution. Proceedings of IAQ '91 - Healthy Buildings, Washington DC, USA, 280-284. Atlanta: American Society of Heating, Refrigerating and Air-Conditioning Engineers, Inc.

Stone, P T (1989). A review of the health aspects of lighting. Contract Report to the Building Research Establishment.

Stridh, G, Andersson, K and Liden, E (1988). A stepwise model for curing buildings with climate problems. Berglund, B and Lindvall, T (Eds), Proceedings of Healthy Buildings 88, Stockholm, Sweden, 3, 247-254. Stockholm: Swedish Council for Building Research.

Strindehag, O, Josefsson, I and Henningson, E (1988). Emission of bacteria from air humidifiers. Berglund, B and Lindvall, T (Eds), Proceedings of Healthy Buildings 88, Stockholm, Sweden, 3, 611-620. Stockholm: Swedish Council for Building Research.

Strobridge, J R and Black, M S (1991). Volatile organic compounds and particle emission rates and predicted air concentrations related to movable partitions and office furniture. Proceedings of IAQ '91 - Healthy Buildings, Washington DC, USA, 292-298. Atlanta: American Society of Heating, Refrigerating and Air-Conditioning Engineers, Inc.

Ström, G, Palmgren, U, Wessén, B, Hellström, B and Kumlin, A (1990). The sick building syndrome. An effect of microbial growth in building constructions? Proceedings of the 5th International Conference on Indoor Air Quality and Climate, Toronto, Canada, 1, 173-178. Ottawa: Canada Mortgage and Housing Corporation.

Sundell, J, Lönnberg, G, Wall, S, Stenberg, B and Zingmark, P A (1990)a. The office illness project in Northern Sweden. Part III: a case-referent study of SBS in relation to building characteristics and ventilation. Proceedings of the 5th International Conference on Indoor Air Quality and Climate, Toronto, Canada, 4, 633-638.

Sundell, J, Wickman, M, Nordvall, L and Pershagen, G (1990)b. Building hygiene and house dust mite infestation. Proceedings of the 5th International Conference on Indoor Air Quality and Climate, Toronto, Canada, 1, 27-29. Ottawa: Canada Mortgage and Housing Corporation.

Sundell, J, Lindvall, T and Stenberg, B (1991). Influence of type of ventilation and outdoor airflow rate on the prevalence of SBS symptoms. Proceedings of IAQ '91 - Healthy Buildings, Washington DC, USA, 85-89. Atlanta: American Society of Heating, Refrigerating and Air-Conditioning Engineers, Inc.

Sundin, B (1982). Formaldehyde emission from wood products; technical experiences and problem solutions from Sweden. 3rd Medical Legal Symposium on Formaldehyde Issues, Washington, D C.

Sutton, P (1991). Noise Control. Kluwer's Handbook of Occupational Hygiene 3, 6.8.

Sverdrup, C, Andersson, K and Andersson, S (1990). A comparative study of indoor climate and human health in 74 day care centers in Malmö, Sweden. Proceedings of the 5th International Conference on Indoor Air Quality and Climate, Toronto, Canada, 1, 651-655. Ottawa: Canada Mortgage and Housing Corporation.

Sverdrup, C F and Nyman, E (1990). A study of microorganisms in the ventilation systems of 12 different buildings in Sweden. Proceedings of the 5th International Conference on Indoor Air Quality and Climate, Toronto, Canada, 4, 583-588. Ottawa: Canada Mortgage and Housing Corporation.

Sykes, J M (1989). Sick building syndrome. Building Services Engineering Research and Technology 10, 1-11.

Tamblyn, R M, Menzies, R I, Comtois, P, Hanley, J, Tamblyn, R T, Farant, J P and Marcotte, P (1991). A comparison of two methods of evaluating the relationship between fungal spores and respiratory symptoms among office workers in mechanically ventilated buildings. Proceedings of IAQ '91 - Healthy Buildings, Washington DC, USA, 136-141. Atlanta: American Society of Heating, Refrigerating and Air-Conditioning Engineers, Inc.

Tancrède, M and Yanagisawa, Y (1990). Volatilization of volatile organic compounds from showers: quantitative assessment and modeling. Proceedings of the 5th International Conference on Indoor Air Quality and Climate, Toronto, Canada, 2, 777-782. Ottawa: Canada Mortgage and Housing Corporation.

Taylor, P R, Dell 'Acqua, B J, Baptiste, M S et al (1984). Illness in an office building with limited fresh air access. Journal of Environmental Health 47, 24-27.

Thomson Laboratories in Association with Building Use Studies and Ove Arup Partnership (1988). Research into the Sick Building Syndrome - Office Environment Survey. Contract report to the Building Research Establishment.

Tong, D (1989). Beyond efficiency. Design Week, January 13.

Tong, D (1991). Sick buildings: what are they and what is their cause? Facilities 9:7, 9-17.

Tsuchiga, Y and Stewart, J B (1990). Volatile organic compounds in the air of Canadian buildings with special reference to wet process photocopying machines. Proceedings of the 5th International Conference on Indoor Air Quality and Climate, Toronto, Canada, 2, 633-638. Ottawa: Canada Mortgage and Housing Corporation.

Tucker, W G (1986). Research overview: sources of indoor air pollutants. Proceedings of the ASHRAE Conference IAQ '86, Managing Indoor Air for Health and Energy Conservation. Atlanta: The American Society of Heating, Refrigerating, and Air-Conditioning Engineers, Inc.

Turiel, I, Hollowell, C D, Biksch, R R, Rudy, J V, Young, R A and Coye, M J (1983). The effects of reduced ventilation on indoor air quality in an office building. Atmospheric Environment 17, 51-64.

Turner, S and Binnie, P W H (1990). An indoor air quality survey of twenty-six Swiss office buildings. Proceedings of the 5th International Conference on Indoor Air Quality and Climate, Toronto, Canada, 4, 27-32. Ottawa: Canada Mortgage and Housing Corporation.

University of Washington Conference (1982). Indoor Air Problems - The Office Environment - A National Epidemic.

Urch, R B, Silverman, F, Corey, P and Shephard, R J (1990). Acute symptom responses to passive cigarette smoke in asthmatic and nonasthmatic individuals. Proceedings of the 5th International Conference on Indoor Air Quality and Climate, Toronto, Canada, 1, 337-342. Ottawa: Canada Mortgage and Housing Corporation.

US Congress, Office of Technology Assessment (1989). Biological Effects of Power Frequency Electric and Magnetic Fields - Background Paper, OTA-BP-E-53. Washington DC: US Government Printing Office.

US Food and Drug Administration (USFDA) (1980). Negative Ion Generators. Washington DC: US Government Printing Office.

Vaculik, F and Plett, E G (1990). Ventilation demand controller. Proceedings of the 5th International Conference on Indoor Air Quality and Climate, Toronto, Canada, 4, 455-460. Ottawa: Canada Mortgage and Housing Corporation.

Valbjørn, O (1991). Indoor air quality management and the sick building syndrome. Roulet, C A (Ed), Proceedings of a Workshop on Indoor Air Quality Management, 11-23, Lausanne. Luxembourg: Commission of the European Communities, Scientific and Technical Communication Unit.

Valbjørn, O and Kousgaard, N (1984). Headache and mucous membrane irritation. An epidemiological study. Berglund, B, Lindvall, T, Sundell, J (Eds), Proceedings of the 3rd International Conference on Indoor Air Quality and Climate, 2, 249-254. Stockholm: Swedish Council for Building Research.

Valbjørn, O, Hagen, H, Kukkonen, E and Sundell, J (1990)a. Indoor Climate and Air Quality Problems, Investigation and Remedy. SBI Report 212. Hørsholm, Denmark: Danish Building Research Institute.

Valbjørn, O, Nielsen, J B, Gravesen, S and Mølhave, L (1990)b. Dust in ventilation ducts. Proceedings of the 5th International Conference on Indoor Air Quality and Climate, Toronto, Canada, 3, 361-364. Ottawa: Canada Mortgage and Housing Corporation.

Vatne, F (1990). Specification of demands. A strategy for better indoor climate. Proceedings of the 5th International Conference on Indoor Air Quality and Climate, Toronto, Canada, 3, 329-333. Ottawa: Canada Mortgage and Housing Corporation.

Ventresca, J A (1991). Operation and maintenance for indoor air quality: implications from energy simulations of increased ventilation. Proceedings of IAQ '91 - Healthy Buildings, Washington DC, USA, 375-378. Atlanta: American Society of Heating, Refrigerating and Air-Conditioning Engineers, Inc.

Virelizier, H, Gaudin, D, Anguenot, F and Aigueperse, J (1990). An assessment of the organic compounds present in domestic aerosols. Proceedings of the 5th International Conference on Indoor Air Quality and Climate, Toronto, Canada, 3, 737-741. Ottawa: Canada Mortgage and Housing Corporation.

Wahlberg, J E and Lidén, C (1988). Is the skin affected by work at visual display terminals? Dermatol Clinics 6, 81-85.

Wallace, L, and Bromberg, S (1984). Plan and preliminary results of the US Environmental Protection Agency's indoor air monitoring program 1982. Berglund, B, Lindvall, T, Sundell, J (Eds), Proceedings of the 3rd International Conference on Indoor Air Quality and Climate, 1, 173-178. Stockholm: Swedish Council for Building Research.

Wallace, L A, Nelson, C J and Dunteman, G (1991). Workplace characteristics associated with health and comfort concerns in three office buildings in Washington DC. Proceedings of IAQ '91 - Healthy Buildings, Washington DC, USA, 56-60. Atlanta: American Society of Heating, Refrigerating and Air-Conditioning Engineers, Inc.

Waller, R A (1984). Case study of a sick building. Berglund, B, Lindvall, T and Sundell, J (Eds), Proceedings of the 3rd International Conference on Indoor Air Quality and Climate, 3, 349-353. Stockholm: Swedish Council for Building Research.

Wallingford, K M and Carpenter, J (1986). Field experience overview: investigating sources of indoor air quality problems in office buildings. Proceedings of IAQ '86: Managing Indoor Air for Health and Energy Conservation, Atlanta, Georgia, 448-453. Atlanta: The American Society of Heating, Refrigerating, and Air-Conditioning Engineers, Inc.

Wanner, H U and Kuhn, M (1984). Indoor air pollution by building materials. Berglund, B, Lindvall, T, Sundell, J (Eds), Proceedings of the 3rd International Conference on Indoor Air Quality and Climate, 3, 35-40. Stockholm: Swedish Council for Building Research.

Warren, P R and Parkins, L M (1984). Window-opening behaviour in office buildings. ASHRAE Transactions 90, 1056-1076.

Weir, B R, Anderson, G E, Hayes, S R and Greenfield, S M (1990). Specification of indoor air model characteristics. Proceedings of the 5th International Conference on Indoor Air Quality and Climate, Toronto, Canada, 4, 231-236. Ottawa: Canada Mortgage and Housing Corporation.

Wenger, J D (1991). Healthy building strategies. Proceedings of IAQ '91 - Healthy Buildings, Washington DC, USA, 3-7. Atlanta: American Society of Heating, Refrigerating and Air-Conditioning Engineers, Inc.

Whorton, M D, Larson, S R (1987). Investigation and work up of tight building syndrome. Journal of Occupational Medicine 29, 142-147.

Wilkes, C R and Small, M J (1990). Air quality model for volatile constituents from indoor uses of water. Proceedings of the 5th International Conference on Indoor Air Quality and Climate, Toronto, Canada, 2, 783-788. Ottawa: Canada Mortgage and Housing Corporation.

Wilkins, A J, Nimmo-Smith, I, Tait, A, Mcmanus, C, Della Salla, S, Tilley, A, Arnold, K, Barrie, M and Scott, S (1984). A neurological basis for visual discomfort. Brain 107, 989-1017.

Wilkins, A J, Nimmo-Smith, I, Slater, A I and Bedocs, L (1989). Fluorescent lighting, headaches and eyestrain. Lighting Research and Technology 21, 11-18.

Wilson, S (1985). Premises of Excellence. London: Building Use Studies.

Wilson, S and Hedge, A (1987). The office environment survey: a study of building sickness. London: Building Use Studies.

Wilson, S, O'Sullivan, P, Hedge, A and Jones, P (1987). Sick building syndrome and environmental conditions: case studies in nine buildings. London: Building Use Studies.

Wolverton, B C, McDonald, R C and Watkins, E A Jr (1984). Foliage plants for removing indoor air pollutants from energy efficient homes. Economic Botany 38, 224-228.

Wolverton, B C, McDonald, R C and Mesick, H H (1985). Foliage plants for the indoor removal of the primary combustion gases carbon dioxide and nitrogen dioxide. Journal of the Mississippi Academy of Sciences 30, Paper 1.

Wolverton, B C and Douglas, W L (1989)a. Sick building syndrome - how do you spell relief? P-L-A-N-T-S. Florida Nurseryman, January.

Wolverton, B C, Johnson, A and Bounds, K (1989)b. Interior landscape plants for indoor pollution abatement. NASA Final Report.

Woods, J E (1989). Cost avoidance and productivity in owning and operating buildings. Occupational Medicine: State of the Art Reviews 4, 753-770.

WHO (World Health Organisation) (1979). Health Aspects Related to Indoor Air Quality: report on a WHO meeting in Bilthoven. Copenhagen: WHO Regional Office for Europe.

WHO (World Health Organisation) (1982). Indoor Air Pollutants: Exposure and Health Effects: report on a WHO meeting in Norlingen. Copenhagen: WHO Regional Office for Europe.

WHO (World Health Organisation) (1986). Indoor Air Quality Research. EURO Reports and Studies No.103. Copenhagen: WHO Regional Office for Europe.

WHO (World Health Organisation) (1987). Indoor Air Quality. Euro reports and studies III. Berlin: World Health Organization.

WHO (World Health Organisation) (1990). Indoor Air Quality: Biological Contaminants, report on a WHO meeting. WHO Regional Publications European Series No.31. Copenhagen: WHO Regional Office for Europe.

Wyatt, T (1991). Chilled beams and displacement ventilation. Building Services 13, 34-36.

Wyon, D P, Asgeirsdottir, T H, Jensen, P K and Fanger, P O (1973). The effects of ambient temperature swings on comfort, performance and behaviour. Archives des Sciences Physiologique 27, 441-458.

Yaglou, C P, Riley, E C and Coggins, D I (1936). Ventilation requirements. ASHRAE Transactions 42, 133-162.

Youle, A (1986). Occupational hygiene problems in office environment: the influence of building services. Annals of Occupational Hygiene 30, 275-287.

Yu, H H S and Raber, R R (1990). Implications of ASHRAE Standard 62-89 on filtration strategies and indoor air quality and energy conservation. Proceedings of the 5th International Conference on Indoor Air Quality and Climate, Toronto, Canada, 3, 121-126. Ottawa: Canada Mortgage and Housing Corporation.

Yu, H H S and Raber, R R (1991). Ventilation and filtration requirements based on Fanger's "Perceived good air quality" and Atlanta: ASHRAE Standard 62-1989. Proceedings of IAQ '91 - Healthy Buildings, Washington DC, USA, 358-362. Atlanta: American Society of Heating, Refrigerating and Air-Conditioning Engineers, Inc.

Zweers, T, Preller, L, Brunekreef, B and Boleij, J S M (1990). Relationships between health and indoor climate complaints and building, workplace, job and personal characteristics. Proceedings of the 5th International Conference on Indoor Air Quality and Climate, Toronto, Canada, 1, 495-500.

Zyla-Wisensale, N H and Stolwijk, J A J (1990). Indoor air quality as a determinant of office worker productivity. Proceedings of the 5th International Conference on Indoor Air Quality and Climate, Toronto, Canada, 1, 249-254. Ottawa: Canada Mortgage and Housing Corporation.

APPENDIX 1. BUILDING SPECIFICATION (From Vatne 1990)

Basic conditions and assumptions for calculations and dimensioning

The intended use of different building areas.
Estimated occupancy, m/persons (WHO 1979).
Internal heat sources, W/m (WHO 1979).
Sunshading devices.
Air pollution from activities and from building materials.
Acceptance criteria.
Accepted outdoor design temperatures.

Ventilation and air quality

Ventilation based on pollution from building materials, l/s/m (WHO 1979).
Ventilation based on occupancy, l/s/person.
Ventilation based on pollution from activities, l/s.
Description and air quantities for exhaust hoods and process-ventilation, l/s.
Resulting total ventilation, l/s/m (WHO 1979).
Possible increase at separate terminal devices.
Possible increase in total ventilation.
Place of air intake in consideration of air outlet and pollution sources.
Filter type for supply air, filter classification.
Description of air balance and intended transferred air.
Heat recovery, type of installation.
Prevention against transfer of polluted air in heat recovery devices.
Intended use of return air.
Use of air humidification, description of system.

Thermal conditions

Operative temperature at outdoor winter and summer design temperatures.
If deviations from recommended comfort requirements may occur, the deviation and duration shall be stated.
Heating method.
Cooling method.

Running and maintenance

Necessary measures and procedures to achieve clean ventilation equipment in new plant. Who is responsible?
Is there necessary access and space for inspections and maintenance of devices and ducts?
Who is responsible for the training of staff responsible for running and maintenance of the installations?
Who is responsible for preparing adequate operating instructions?
Who is responsible for balancing and adjusting the installations?
Does the contract include measurements and control verifying the thermal design figures?

APPENDIX 2. ENVIRONMENTAL FACTORS SUSPECTED OF CAUSING SBS SYMPTOMS, PROBABLE INDICATORS AND EXAMPLES OF POSSIBLE REMEDIES (from Valbjørn et al 1990)

HUMAN BIOEFFLUENTS

Indicator. Carbon dioxide concentration > 1000 ppm; odour.

Remedy. Check and adjust operation and maintenance of ventilation system; adjust to new requirements; increase the outdoor air rate; give airing instructions.

X HIGH TEMPERATURE

Indicator. Complaints or temperatures above 23-24°C.

Remedy. Reduce heat load (from lighting, sun and equipment); control and maintenance of heating plant.

RADIATION OF HEAT AGAINST THE HEAD

Indicator. Ceiling heating (surface temperature above 35°C).

Remedy. Change incandescent lamp to fluorescent lighting with high-frequency ballasts; replace ceiling heating with radiator.

MOULD SPORES

Indicator. Visible mould; odour; water damage; condensation; many plants (spore content in the air).

Remedy. Wash away the mould; reduce humidity load; deal with leaks in roofs and walls; increase ventilation.

MICRO-ORGANISMS FROM HUMIDIFIERS

Indicator. Stagnant water in humidifier (e.g. night/weekends); visible growth.

Remedy. Clean humidifier; check maintenance and operation; switch off or change to a steam model.

ORGANIC DUST IN CARPETING

Indicator. Visible dirt; carpet 5-10 years old not cleaned annually; analysis of vacuum-cleaned dust.

Remedy. Try thorough cleaning; change to 'hard' flooring.

LEVELLING COMPOUND IN THE FLOOR STRUCTURE

Indicator. Casein filler; discoloured floors; odour.

Remedy. Remove casein and dry out construction.

X HIGH DUST CONCENTRATION IN AIR

Indicator. Visible dust on floors, shelves or in the air; much paper handling and high activity.

Remedy. Improve cleaning; improve filtration.

X TOBACCO SMOKE

Indicator. Odour; carbon monoxide; questionnaire.

Remedy. Check smoking in proportion to odour transfer between rooms (stop recirculation); check ventilation rate; avoid smoking.